MASTERING THE

GEORGIA 5TH GRADE CRCT

IN SCIENCE

Written to GPS 2006 Standards

Liz Thompson
Michelle Gunter

American Book Company
PO Box 2638
Woodstock, GA 30188-1383
Toll Free: 1 (888) 264-5877 Phone: (770) 928-2834
Fax: (770) 928-7483 Toll Free Fax: 1 (866) 827-3240
Web site: www.americanbookcompany.com

ACKNOWLEDGEMENTS

The authors would like to gratefully acknowledge the formatting and technical contributions of Becky Wright.

We also want to thank Mary Stoddard and Eric Field for their expertise in developing the graphics for this book.

A special thanks to Marsha Torrens for her editing assistance.

PO Box 2638
Woodstock, GA 30188-1318

Printed in the United States of America

10/07

Table of Contents

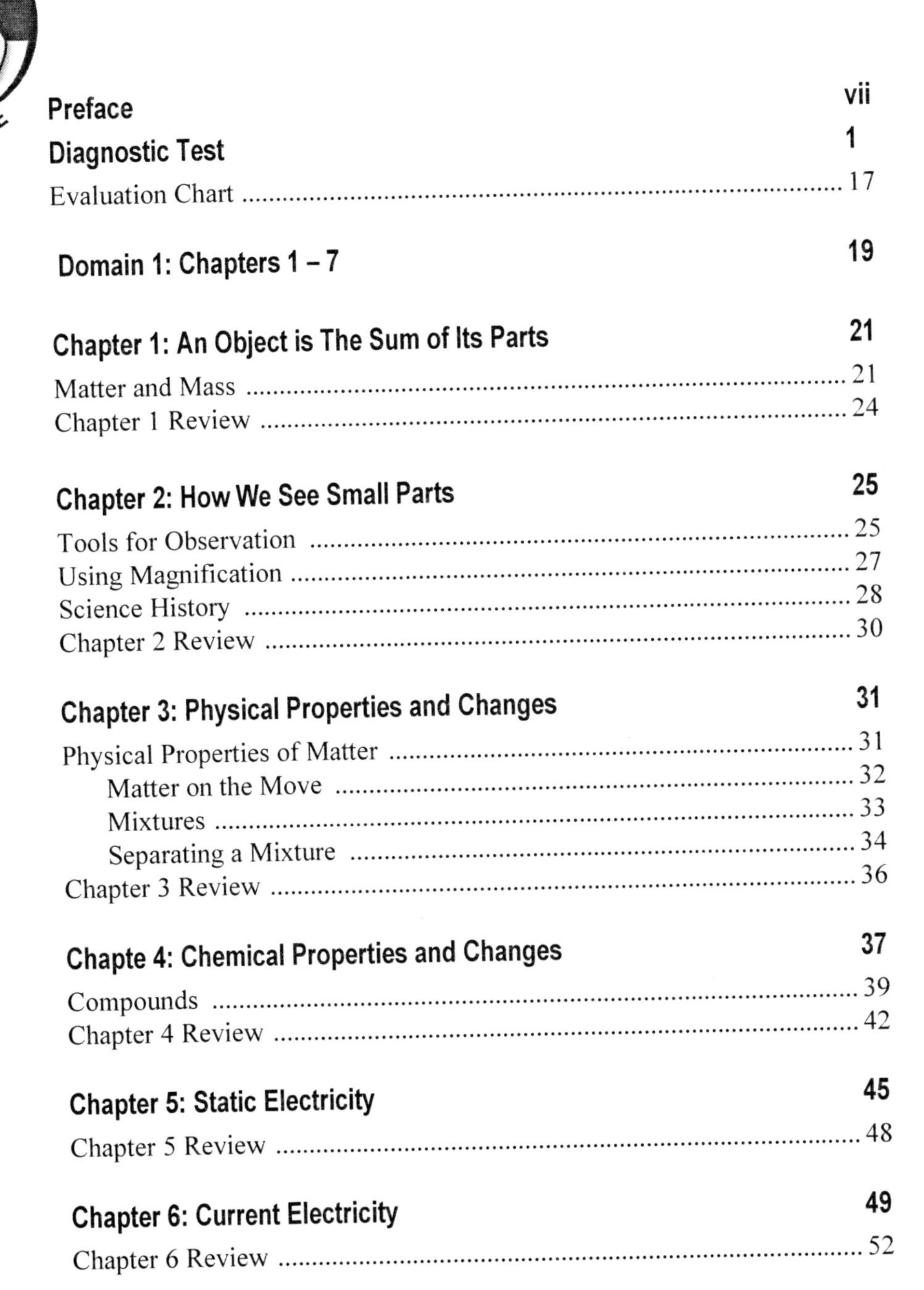

Preface vii

Diagnostic Test 1

Evaluation Chart 17

Domain 1: Chapters 1 – 7 19

Chapter 1: An Object is The Sum of Its Parts 21

Matter and Mass 21
Chapter 1 Review 24

Chapter 2: How We See Small Parts 25

Tools for Observation 25
Using Magnification 27
Science History 28
Chapter 2 Review 30

Chapter 3: Physical Properties and Changes 31

Physical Properties of Matter 31
- Matter on the Move 32
- Mixtures 33
- Separating a Mixture 34

Chapter 3 Review 36

Chapte 4: Chemical Properties and Changes 37

Compounds 39
Chapter 4 Review 42

Chapter 5: Static Electricity 45

Chapter 5 Review 48

Chapter 6: Current Electricity 49

Chapter 6 Review 52

Chapter 7: Magnetism 53
Electromagnetic Force and Fields 55
Electromagnets 55
Electromagnets vs. Bar Magnets 58
Chapter 7 Review 60

Domain 1 Review 61

Domain 2: Chapters 8 – 13 63

Chapter 8: Classification 65
Chapter 8 Review 68

Chapte 9: Animal Groups 69
Chapter 9 Review 74

Chapte 10: Plant Groups 75
Chapter 10 Review 78

Chapte 11: Cells 79
Parts of a Cell 79
Science History 79
Plant Cell 80
Animal Cell 81
Cell Differences in Multicellular Organisms 82
Multicellular Cells vs. Unicellular Cells 82
Chapter 11 Review 84

Chapter 12: I Like Your Genes 85
Chapter 12 Review 87

Chapter 13: Surrounded by Germs 89
Chapter 13 Review 92

Domain 2 Review 95

Domain 3: Chapters 14 – 16 97

Chapter 14: Plate Tectonics 99
The Earth's Layers 99
Plate Tectonics 100
Continental Drift and Seafloor Spreading 101
Constructive and Destructive Processes 102
Chapter 14 Review 104

Chapter 15: Rocks and the Rock Cycle 105

Rocks and the Rock Cycle 105
Weathering 105
Erosion 106
Constructive and Destructive Processes 107
Volcanic Activity 108
    The Rock Cycle 109
Chapter 15 Review 110

Chapter 16: Managing Earth's Changes 113

Preparing for the Possibilities 113
    Flood Control 113
    Beach Reclamation 116
    Seismological Studies 117
Chapter 16 Review 120

Domain 3 Review 123

Post Test 1 125

Post Test 2 139

Appendix A 151

Preface

The Georgia 5th Grade CRCT Test in Science will help students who are learning or reviewing material for the Georgia test that is now required for each gateway or benchmark course. **The materials in this book are based on the Georgia Performance Standards as published by the Georgia Department of Education.**

This book contains several sections. These sections are as follows: 1) General information about the book; 2) A Diagnostic Test and Evaluation Chart; 3) Domains/Chapters that teach the concepts and skills to improve readiness for Georgia 5th grade CRCT test in Science; 4) Two Post/Practice Tests. Answers to the tests and exercises are in a separate manual. The answer manual also contains a Chart of Standards for teachers to make a more precise diagnosis of student needs and assignments.

We welcome comments and suggestions about the book. Please contact us at

American Book Company
PO Box 2638
Woodstock, GA 30188-1383

Toll Free: 1 (888) 264-5877
Phone: (770) 928-2834
Fax: (770) 928-7483
Web site: www.americanbookcompany.com

About the Authors

Liz A. Thompson holds a B.S. in Chemistry and an M.S. in Analytical Chemistry, both from the Georgia Institute of Technology. Research conducted as both an undergraduate and graduate student focused on the creation and fabrication of sensors based on conducting polymers and biomolecules. Post graduate experience includes work in radioanalytical chemistry. Her publications include several articles in respected scientific journals, as well as partial authorship of the textbook *Radioanalytical Chemistry* (2007). At every educational level, Mrs. Thompson has enjoyed teaching, tutoring and mentoring students in the study of science.

Michelle Gunter graduated from Kennesaw State University in Kennesaw, Georgia with a B.S. in Secondary Biology Education. She is a certified teacher in the field of Biology in the state of Georgia. She has three years experience in high school science classrooms. She has nine years experience in biology and biological systems. She has won awards for her research in the field of aquatic toxicology. Mrs. Gunter enjoys teaching students of all ages, the wonders of the natural world.

5th Grade CRCT Science Diagnostic Test

1. How do tectonic plates move? S5E1a, b

   A They roll.

   B They float.

   C They bounce.

   D They jump.

2. Which pair of lists below groups animals into amphibians and reptiles? S5L1a

   A

human butterfly frog	lizard turtle snake

   B

salamander frog newt	iguana turtle snake

   C

carrot dog snake	turtle frog ladybug

   D

iguana dove parrot	blue whale shark squid

3. Which piece of scientific equipment would you use to see human cells? S5L3a

   A telescope

   B microscope

   C microwave

   D computer

4. In 1960, Harry Hess described his idea that the seafloor was spreading from cracks in the ocean floor. The magma oozes out of the cracks, then hardens into rock. What does Mr. Hess's idea tell you about the age of the rock found 1,000 km from the crack? S5E1a

   A The rock farther away from the crack is younger than the rock closer to the crack.

   B The rock is the same age no matter how far away from the crack.

   C The rock farther away from the crack is composed of different material than rock closer to the crack.

   D The rock farther away from the crack is older than the rock closer to the crack.

5. Which of the following objects could NOT be picked up with an electromagnet? S5P3d

   A staples

   B aluminum cans

   C rubber bands

   D keys

6. What type of disease can you get from fungi? S5L4b

   A measles

   B chicken pox

   C food poisoning

   D athlete's foot

7. A destructive process S5E1b

   A builds up landforms.

   B breaks down landforms.

   C makes volcanoes.

   D moves tectonic plates.

Go On

Jackson uses four different Legos® to build a tower. Each Lego® has a different mass. Use the following graph to answer questions 8 – 9.

Mass (in grams)
of the Legos® in a Tower

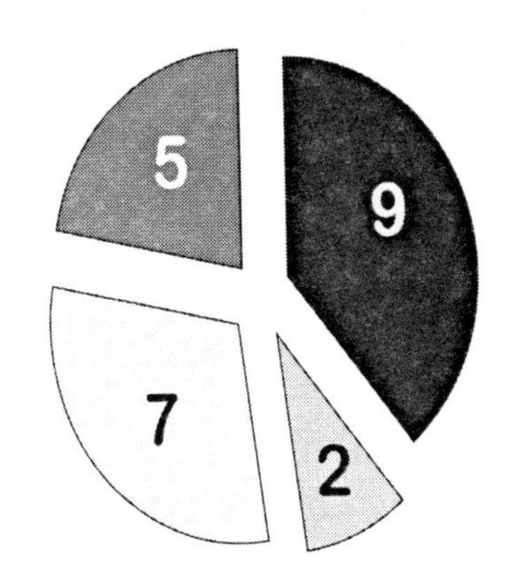

8. What is the mass of Jackson's tower? S5P1a

   A 16 grams

   B 18 grams

   C 21 grams

   D 23 grams

9. Jackson wants to greatly reduce the mass of the tower. Which one Lego® should Jackson remove? S5P1a

   A the 9 gram Lego®

   B the 7 gram Lego®

   C the 5 gram Lego®

   D the 2 gram Lego®

The following is the procedure Tamika followed for her science lab. Use this information to answer question #10.

- Step 1: Add water to glass jar.
- Step 2: Heat water until boiling.
- Step 3: Add 4 drops blue food coloring to water.
- Step 4: Stir.
- Step 5: Allow to cool.

10. During which steps did Tamika cause a chemical change? S5P2b

    A Steps 2 and 5

    B Steps 3 and 4

    C Steps 2 and 3

    D No chemical change occurred during the lab.

11. The giant sequoia is one of the world's tallest trees, with average heights of 50 – 95 meters. What allows these trees to grow so tall? S5L1b

    A leaves

    B spine

    C vascular tissue

    D non-vascular tissue

12. Which of the following would NOT erode soil? S5E1b

A a rainstorm

B a river

C a tornado

D grass

13. What is the BEST description of static electricity? S5P3a

A surface charge that can't move

B mobile surface charge

C current in a wire

D an electrical surface

14. The ability to talk is BEST described as S5L2a

A genetic trait.

B learned behavior.

C DNA behavior.

D result of protein production.

15. A liquid sample is given to Amy in lab. Before the lab, it was clear and bubbly. After the lab, it was clear and not bubbly. Based only on this information, what is a correct conclusion? S5P2c

A The liquid sample has not changed.

B The liquid sample has chemically changed.

C The amount of gas mixed into the liquid has decreased.

D The amount of solid mixed into the liquid has decreased.

16. Which of the following physical properties is qualitative — that is, described using words instead of a number? S5P2a

A length

B color

C mass

D boiling point

Go On

17. Providence Canyon is located in Stewart County, GA. This canyon was formed partially by natural processes. But a lot of the erosion that formed the canyon was not natural. It was caused by poor farming practices in the 1800s. What conclusion can you draw from this information? S5E1b, c

A All canyons are formed by poor farming practices.

B Erosion can be natural or it can be caused by humans.

C Erosion usually doesn't form canyons like Providence Canyon.

D Without farmers, there would be no Providence Canyon.

18. Which of the following materials is the MOST electrically conducting? S5P3c

A wood

B copper

C water

D plastic wrap

19. A mosquito bites a human. Several days later the human begins to get sick. What type of microbe is LIKELY causing the illness? S5L4b

A bacteria

B fungi

C protists

D virus

20. Electric current is BEST described as the S5P3b, c

A flow of electric charge through an insulator.

B flow of electrical charge through a conductor.

C the resistance to electrical charge caused by a resistor.

D the creation of electrical charge in a power source.

Use the following information to answer question #21.

Mallorie spent a week with her new puppy, Tipper. The table below shows what Mallorie and Tipper did each day.

Day of the Week	Behavior
Monday	Walking on a leash
Tuesday	Running around the backyard
Wednesday	Swimming in the pond
Thursday	Fetching the paper
Friday	Barking at birds

21. Which of Tipper's behaviors were learned? S5L2b

A walking on a leash and barking at birds

B walking on a leash and fetching the paper

C swimming in the pond and running around the backyard

D swimming in the pond and barking at birds

Go On

22. What do genes do? S5L1b

A They tell the cell what proteins to make.

B They tell the cell to store energy.

C They make your hair long.

D They make you look good for your friends.

23. Where would be the BEST place to put a levee? S5E1c

A in the center of a river

B around a low-lying lake

C in a storm drain

D on top of a dam

24. What piece of a microscope magnifies the object you are looking at? S5P1b

A the light source

B the adjustment knobs

C the nosepiece

D the objective

25. Which of the following makes a mixture? S5P2a

A tearing up a sheet of paper

B stirring ice cubes into water

C pouring instant cocoa into hot milk

D turning on more hot water in your shower

26. Where does your DNA come from? S5L2b

A It all comes from your mother.

B It all comes from your father.

C Half of it comes from your mother and half of it from your father.

D Half of it comes from your environment and half of it comes from your parents.

Use the following image to answer question #27.

27. What would happen if you removed the electrodes from the salt water? S5P3b

A The light bulb would get brighter.

B The water would evaporate.

C The light bulb would go out.

D The salt water would light up the air around it.

Go On

28. A solid material is heated over a flame. When it reaches its melting point, what will happen? S5P2b

A It will turn into a liquid.

B It will ignite.

C It will disappear.

D It will become magnetic.

29. A bar magnet is a permanent magnet and is surrounded by magnetic field lines. Describe the direction of its magnetic field lines. S5P3d

A from the north pole of the magnet to its south pole

B from the south pole of the magnet to its north pole

C upwards from a central point between the poles

D directly outwards from the south pole

30. Which example below is an inherited trait? S5L2a

A tying your shoes

B reading a book

C the shape of your mouth

D the language you speak

31. Which characteristic below is ONLY true of reptiles? S5L1a

A lays eggs

B has hair

C is endothermic

D has dry, scaly skin

32. What is stored in the nucleus? S5L3b

A chloroplasts

B cytoplasm

C water

D DNA

33. Dante washed the family car in the driveway. It was a cool November day. Three hours later, the driveway was dry. What happened? S5P2b

A The water evaporated.

B The water condensed.

C The water was mostly soaked up by the driveway.

D The water chemically changed into air.

34. A tsunami is a large wave of water, generated deep in the ocean. What process often generates a tsunami? S5E1b

A seafloor spreading

B a deep sea trench

C an earthquake

D an eclipse

Go On

35. Identify the cell wall in the diagram below. S5L3b

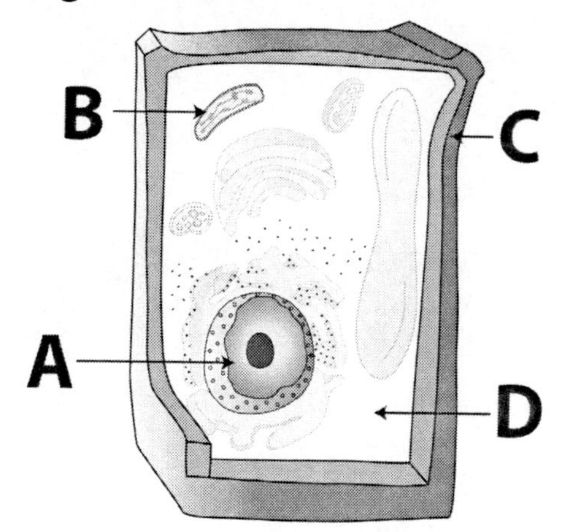

STOP

36. What structure results from similar cells grouping together? S5L3b

A an organ

B tissue

C an organ system

D an organism

37. Which of the following is a destructive process? S5E1b

A sand dune formation

B sand dune erosion

C sand dune preservation

D sand dune protection

38. Describe what occurs in this series of pictures. S5P2a, c

A The paper undergoes two physical changes.

B The paper undergoes a physical change followed by a chemical change.

C The paper undergoes a chemical change followed by a physical change.

D The paper undergoes two chemical changes.

39. The cells in your body S5L3c

A are the exact same as an amoeba cell.

B are the exact same as a bacterial cell.

C are similar to an amoeba cell.

D are similar to a bacterial cell.

40. Why do city planners maintain the edges of waterways with grass and landscaping? S5E1b

A to increase water flow

B to increase erosion

C to decrease water flow

D to decrease erosion

41. How does the human body fight off a bacterial infection? S5L4b

A with the immune system

B with DNA

C with a fever

D with your fingers

42. Which of the following materials would allow the greatest build up of static electricity? S5P3a

A silver

B glass

C aluminum

D copper

Go On

43. If the two animals shown here had babies together, what would the puppies look like? S5L2b

A The puppies would look like a poodle.

B The puppies would look like a bulldog.

C The puppies would have no ears.

D The puppies would look like a mixture of a poodle and a bull-dog.

44. Which of the following contributes the least to the destruction of landforms? S5E1b

A wind

B sunlight

C rain

D lichen

45. Which item below is the part of a plant cell that captures energy? S5L3b

A nucleus

B chloroplasts

C cell membrane

D cell wall

46. The ancient supercontinent Pangaea broke apart about 200 million years ago. What was this process called? S5E1b

A continuous drift

B continental drift

C destructive tectonics

D continental tectonics

47. A volcano erupting can be considered a constructive process for what reason? S5E1a

A It makes the ground very hot.

B It makes the ground steeper.

C It makes the ground higher.

D It makes the ground lower.

48. What characteristic listed below places plants in a group separate from animals? S5L1a, b

A They are multicellular.

B They are vertebrates.

C They make their own food.

D They reproduce.

49. Which of the following is NOT directly caused by seismic activity? S5E1a, b, c

A erosion

B earthquakes

C volcanoes

D tsunamis

Go On

50. What are vaccines? S5L4b

A They are similar to genes because they store energy.

B They kill viruses using antibacterial medicine.

C They prevent viral infections using your immunity.

D They are widespread infections.

Use the image to answer question 51.

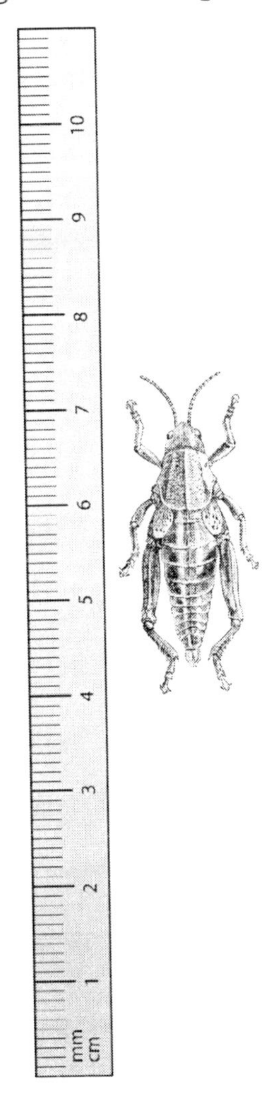

51. Which magnification level would make this grasshopper appear to be 30-35 cm long? S5L3a

A 2x

B 5x

C 10x

D 15x

52. A hair dryer is plugged into an electrical circuit. Which circuit element is the hair dryer? S5P3b

A a closed switch

B an open switch

C a resistor

D a power source

53. How was the color of your hair determined? S5L2a

A from your DNA

B from your behavior

C from your home

D from the food you eat

Several different types of microscope are used by modern biologists. With these technologies, they can view very tiny objects. Use the following table to answer question #54.

Smallest Size of Specimen	Type of Microscope Needed
The surface of the skin on your face	Simple magnifying glass
The surface of small particles: dust, pollen, salt crystals	Light microscope
More detail than the light microscope on even smaller objects	Electron microscope
Individual molecules and atoms	Atomic force microscope

54. What type of magnification do you think you would use to examine the outside of a flea? S5P1b

A simple magnifying glass

B light microscope

C electron microscope

D atomic force microscope

Go On

55. Describe a plastic-covered electrical wire. S5P3c, b

A an insulator covered by a conductor

B a conductor covered by an insulator

C a power source covered by a resistor

D a resistor covered by a power source

56. Alberto placed his pet ferret on the household scale. The ferret had a mass of 130 kilograms. Which of the following masses was NOT a component of the total mass of the ferret? S5P1a

A his toenails

B his tail

C the air around him

D the air in his lungs

57. Which microorganisms listed below can be harmful? S5L4b

A bacteria

B fungi

C protists

D all of the above

58. In the following concept map, some items are missing. Which of the following is NOT an item that could appear as a physical property? S5P2a

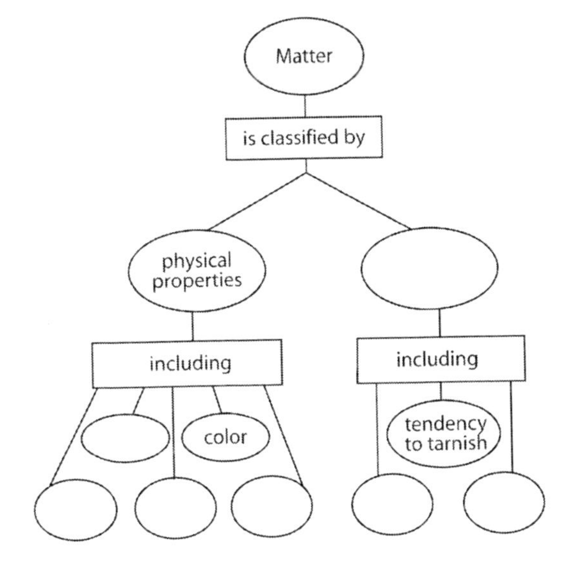

A smell

B flammability

C shape

D taste

59. A good example of static electricity is S5P3a

A the current flowing through an electrical circuit.

B the current that flows through car jumper cables.

C a battery.

D lightening.

60. Which characteristic below is ONLY true of birds? S5L1a

    A has hair

    B lays eggs

    C has feathers

    D makes milk

61. The main difference between a bar magnet and an electromagnet is S5P3d

    A their size.

    B the fact that bar magnets work without electrical current.

    C the fact that bar magnets require an electrical current.

    D the direction of their magnetic fields.

62. Mosses and liverworts are two examples of a plant. Which common characteristic do they share? S5L1b

    A a spine

    B the ability to bear fruit

    C vascular tissue

    D non-vascular tissue

63. Which Georgia landform will erode the MOST sediment? S5E1b

    A the Okefenokee Swamp

    B Brasstown Bald

    C the Chattahoochee River

    D the Georgia Barrier Islands

64. Which cell part is mostly water? S5L3b

    A mitochondria

    B nucleus

    C cytoplasm

    D chloroplasts

65. Which of the following is comparable to the speed at which an electrical charge moves? S5P3b

    A a 5th grader running at full speed

    B a cheetah running at full speed

    C a BMW™ traveling at 75 miles per hour

    D faster than a BMW™ traveling at 75 miles per hour

66. Which pair of lists below groups animals into vertebrates and invertebrates? S5L1a

A

snake	carrot
goldfish	orchid
finch	pine

B

iguana	sand dollar
ant	bacteria
tomato	alga

C

spider	elephant
jellyfish	bass
earthworm	wolf

D

tortoise	squid
redwood	jewel beetle
killer whale	wasp

67. Name one thing that will increase the magnetism of an electromagnet? S5P3d

A attach a bar magnet to it

B increase the current running through it

C decrease the current running through it

D drop or heat it

68. A 5 gram sample of iron filings is placed in each of four containers. Use the observations recorded in the following table to choose the container in which a chemical change occurred. S5P2c

Container	Contents	Observations
1	4 mL of a liquid	The container warms.
2	4 grams of a solid	Filings stick to the solid.
3	5 grams of a solid	A mixture forms.
4	5 grams of a liquid	Filings sink to the bottom.

A Container 1

B Container 2

C Container 3

D Container 4

69. Flooding is more common in low lying areas. Which of the following cities will likely have the MOST levees to control flooding? S5E1c

A Atlanta, Georgia: 300 meters above sea level

B Miami, Florida: 2 meters above sea level

C Salt Lake City, Utah: 1288 meters above sea level

D Denver, Colorado: 1610 meters above sea level

70. Which of the following items changes electrical current into heat? S5P3b

A a fire

B a kerosene lamp

C a heating pad

D a telephone

STOP

EVALUATION CHART

GEORGIA 5TH GRADE CRCT IN SCIENCE DIAGNOSTIC TEST

Directions: On the following chart, circle the question numbers that you answered incorrectly, and evaluate the results. Then turn to the appropriate topics (listed by chapters), read the explanations, and complete the exercises. Review other chapters as needed. Finally, complete the Post Tests to prepare for the Georgia 5th Grade CRCT in Science.

Chapters	Question Numbers
Chapter 1: An Object is the Sum of Its Parts	8, 9, 56
Chapter 2: How We See Small Parts	24, 54
Chapter 3: Physical Properties and Changes	10, 16, 25, 28, 38, 58
Chapter 4: Chemical Properties and Changes	15, 38, 68
Chapter 5: Static Electricity	13, 18, 20, 42, 55, 59
Chapter 6: Current Electricity	20, 27, 55, 65, 70
Chapter 7: Magnetism	5, 29, 61, 67
Chapter 8: Classification	2, 48
Chapter 9: Animal Groups	2, 31, 48, 60, 66
Chapter 10: Plant Groups	11, 22, 48, 62
Chapter 11: Parts of a Cell	3, 32, 35, 36, 39, 45, 51, 52, 64
Chapter 12: I Like Your Genes	14, 21, 26, 30, 33, 43, 53
Chapter 13: Surrounded by Germs	6, 19, 41, 50, 57
Chapter 14: Plate Tectonics	1, 4, 7, 12, 17, 34, 37, 40,44, 46, 47, 49, 63
Chapter 15: Rocks and the Rock Cycle	1, 4, 7, 12, 17, 34, 37, 40, 44, 46, 47, 49, 63
Chapter 16: Managing Earth's Changes	17, 23, 49, 69

Go On

Go On

Domain 1
Chapters 1 – 7

Chapter 1: An Object is the Sum of Its Parts

S5P1: Students will verify that an object is the sum of its parts.

a. Demonstrate that the mass of an object is equal to the sum of its parts by manipulating and measuring different objects made of various parts.

Chapter 2: How We See the Small Parts

S5P1: Students will verify that an object is the sum of its parts.

b. Investigate how common items have parts that are too small to be seen without magnification.

Chapter 3: Physical Properties and Changes

S5P2: Students will explain the difference between a physical change and a chemical change.

a. Investigate physical changes by separating mixtures and manipulating (cutting, tearing, folding) paper to demonstrate examples of physical change.

b. Recognize that the changes in state of water (water vapor/steam, liquid, ice) are due to temperature differences and are examples of physical change.

Chapter 4: Chemical Properties and Changes

S5P2: Students will explain the difference between a physical change and a chemical change.

c. Investigate the properties of a substance before, during and after a chemical reaction to find evidence of change.

Chapter 5: Static Electricity

S5P3: Students will investigate the electricity, magnetism, and their relationship.

a. Investigate static electricity.

c. Investigate common materials to determine if they are insulators or conductors of electricity.

Chapter 6: Current Electricity

S5P3: Students will investigate the electricity, magnetism, and their relationship.

b. Determine the necessary components for completing an electric circuit.

Chapter 7: Magnetism

S5P3: Students will investigate the electricity, magnetism, and their relationship.

d. Compare a bar magnet to an electromagnet.

Chapter 1
An Object is The Sum of Its Parts

MATTER AND MASS

Mass describes how heavy something is. **Matter** is a word that describes anything that has mass and takes up space. A chair is matter. A pencil is matter. A hair and a speck of dust are also matter, even though they have very little mass and take up very little space. Here is something interesting though: each of these things is made up of *other* things.

Think about the chair for a minute. Inside of a chair, there is a wooden frame, held together by metal nails, covered with cloth and stuffed with some kind of batting (either natural cotton or man-made polyester). It probably has a seat cushion made of foam also.

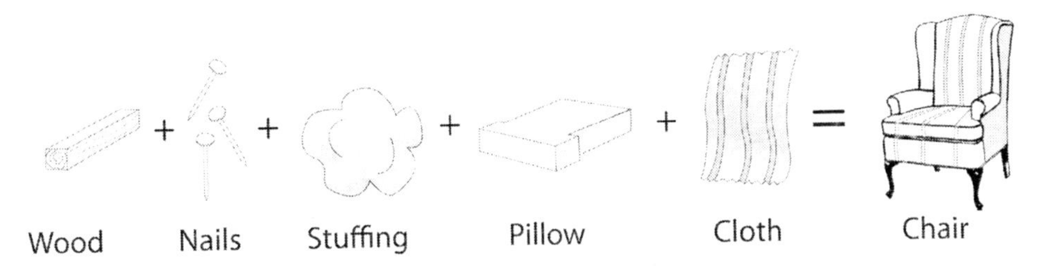

Figure 1.1 Parts of a Chair are All Matter

The wood, the nails, the cloth, the pillow and the stuffing are all matter, too. They have mass and they take up space. Each of them has its own properties. **Properties** are the things that describe an object. Table 1.1 contains some properties of matter.

Table 1.1 Qualitative Properties of Matter

Property	Some choices
color	yellow, purple, brown
odor	fishy, flowery, smoky
texture	rough, smooth, slippery

The properties above can be described using words. They are **qualitative descriptions.** Other properties are defined by numbers. They can be measured. You can say that using the number quantifies the property: it is a **quantitative description**.

Whenever a property is described by a number, it also has a **unit**. **Metric units** are used in science. More properties and their units are given in Table 1.2.

Table 1.2 Quantitative Properties of Matter

Property	Unit
length	meter (m)
mass	gram (g)
temperature	degrees Celsius (°C) or Kelvin (K)

So, let's say that we want to describe the chair in Figure 1.1. We can start out with qualitative descriptions. We can say that it is black and white, and that the cloth covering is smooth. We could say the chair smells fresh and new. We can also use quantitative descriptions. We can measure its height and say that it stands about 1½ meters high. We could say that it costs $450. We could also take the chair's temperature, but that would be silly.

Hey, is the chair easy to lift? We can decide that if we know its mass. If the chair has a mass of 50 kilograms (110 pounds), then it is pretty heavy. If the chair has a mass of 4.5 kilograms (10 pounds) then it is much lighter. We will need to put the chair on a scale to measure its mass (Figure 1.2).

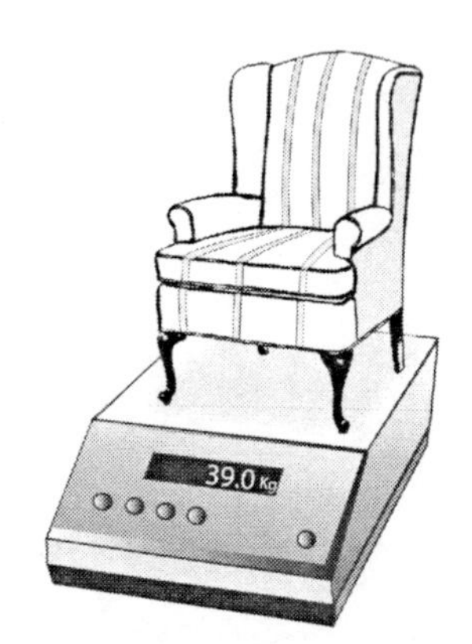

Figure 1.2 Massing the Chair

OK, the chair has a mass of 39 kilograms (86 pounds). So, it may be too heavy for you to lift, but an adult probably could pick it up, or at least slide it across the floor. Good information to have. But wait! Where does that mass come from? In Figure 1.1, you saw all of the parts that make up the chair. Which part gives the chair its mass? Well…they all do.

Wood	29 kilograms
Nails	1 kilogram
Stuffing	1 kilogram
Pillow	3 kilograms
Cloth	5 kilograms
	39 kilograms

Figure 1.3 Total Mass of the Chair

The mass of the chair is exactly equal to the mass of the things that make up the chair. In fact, the mass of any piece of matter is equal to the combined mass of all the things that it contains. This is called the **Law of Conservation of Matter**. It states that the mass in the chair cannot be created or destroyed, though it can be rearranged. Use this information to do the Activity.

Activity

The key contains the mass of each size LEGO® block. Use it to determine the mass of the LEGO® tower.

D D D D D D
C C C
C C C
C C C
C B
A

D = 1 g
C = 2 g
B = 4 g
A = 6 g

Use the key to determine the mass of the bottle of water.

Cap = 0.5 grams

Label = 0.25 grams

Bottle = 1.25 grams

Water = 48 grams

Chapter 1 Review

1. Which of the following properties can be quantified (described by a number)?
   A. color
   B. texture
   C. temperature
   D. odor

2. Which of the following is a metric unit?
   A. pound
   B. gram
   C. feet
   D. inch

3. Which of the following is not matter?
   A. a tissue
   B. an eyelash
   C. a cricket
   D. an idea

Use the following information to answer questions 4 – 5.

Francesca harvests an ear of corn from the garden. She removes the husk, which has a mass of 25 grams. She removes the silk, which has a mass of 5 grams. She is left with the ear of corn on the cob, which has a mass of 155 grams.

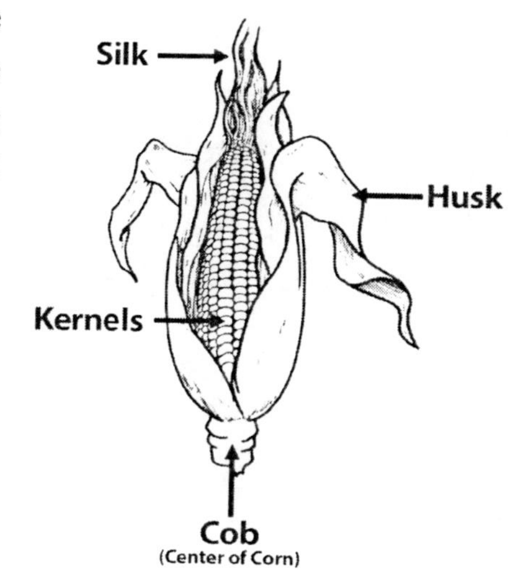

4. What was total mass of the ear of corn when Francesca took it from the garden?
   A. 30 g
   B. 150 g
   C. 180 g
   D. 185 g

5. There are 120 kernels of corn attached to the corncob. Each kernel has a mass of 1 gram. If Francesca eats all the kernels, what is the mass of the cob?
   A. 25 g
   B. 30 g
   C. 35 g
   D. 40 g

Challenge Question

Name four other properties of matter.

Chapter 2
How We See Small Parts

GEORGIA 5 CRCT SCIENCE

Tools for Observation

What is this laboratory tool used for? Correct! It is used to make things larger. This is called **magnification**. We use a handheld **magnifying glass** to magnify normal sized things, simply to see them better. A magnifying glass will increase the size of the object that you are looking at (observing) by about 2 to 4 times (2x – 4x). For instance, in Figure 2.2 we use a magnifying glass to better see the working gears of a clock. We can see the gears without the magnifying glass, but using it makes the gears appear larger. Larger things are easier to observe.

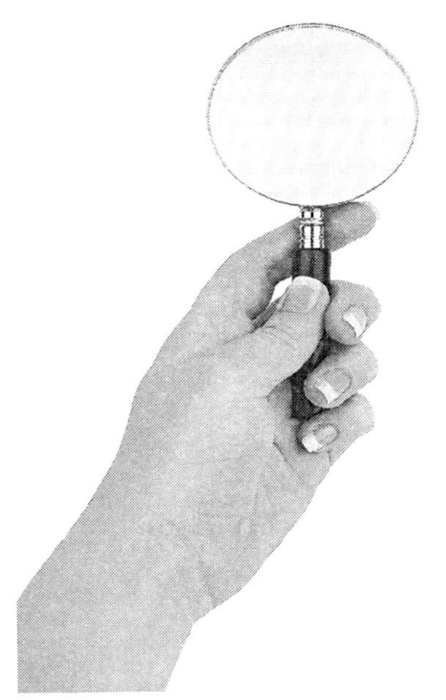

Figure 2.1 A Handheld Magnifying Glass

Figure 2.2 Using a Magnifying Glass to See Clock Gears

What if you want to look even closer? Well, then you need **a microscope**. Microscopes can magnify things a little bit or a whole lot. It all depends on your *objective*. Actually, that is a little joke. An **objective** is a goal, but it is also the name for the piece of a microscope that provides magnification. Take a look at Figure 2.3.

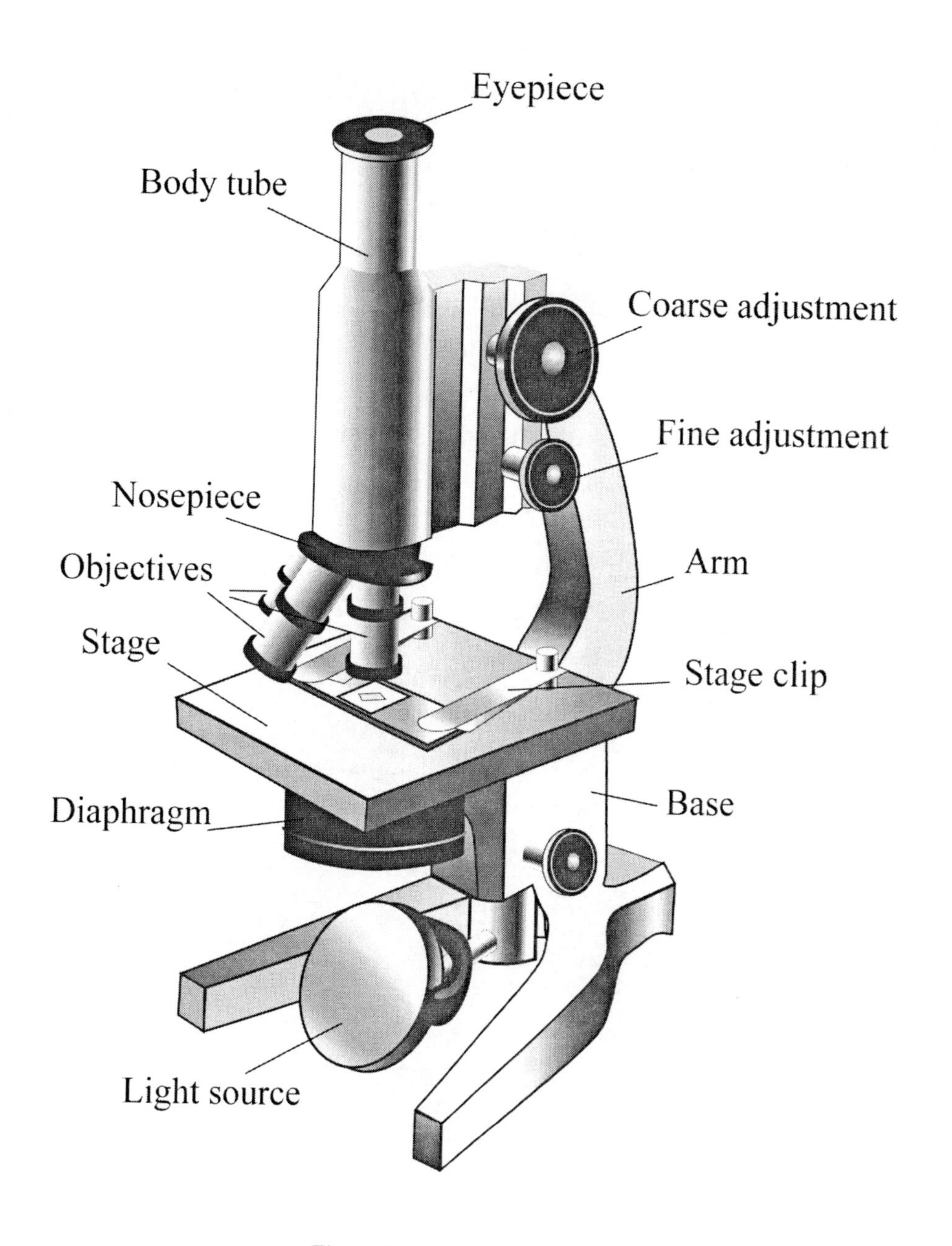

Figure 2.3 A Light Microscope

Figure 2.3 shows an **optical microscope**. This is also called a **light microscope**. It uses light (from the light source) and optical lenses (in the objectives) to magnify objects. The objectives of a regular student microscope can usually magnify things from 4x to 100x their size. More advanced light microscopes can magnify up to about 1000x. You can see a whole range of objects with that much magnification. Using a 4x objective, you could magnify a 3 centimeter (cm) paper clip so that it appeared to be 12 cm. (It may actually be even larger, because the eyepiece of the microscope also magnifies. Let's stick to the objectives right now, though.)

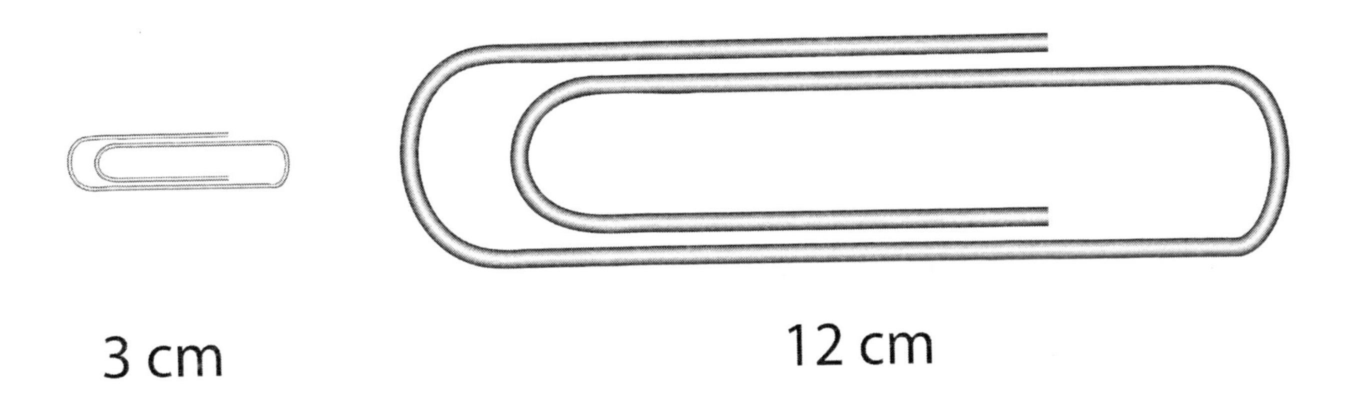

Figure 2.4 Magnifying a Paper Clip

Of course, in order to see the paper clip clearly, you will have to make sure that it is in focus. An image is **in focus** when it is sharp and clear. **Unfocused** images are blurry. To focus the image, turn the adjustment knobs. Turning the **coarse adjustment knob** will get the image in focus, but you will need to turn the **fine adjustment knob** to see the sharpest image.

Now that it is focused, what if you want more magnification? Turn the **nosepiece** of the microscope to get a different objective! Using a 100x objective, your paper clip would appear 300 cm long. That is about 3 feet!

When things appear that large, you can see them really well. You can examine their surfaces to determine texture — is it rough or smooth, does it have scratches? Does it look man-made or natural? You can observe color — is it all the same color (uniform) or is it patterned? You can make all kinds of observations. Just open your eyes and look at the big picture!

USING MAGNIFICATION

So, we can make things larger using magnification. Why is that useful? One reason is that humans are very visual creatures. That means that we often need to see things in order to understand them. And there are a lot of things that humans are trying to understand.

Have you ever seen the PBS series *The Magic School Bus* ™? The wonderful thing about that show is that it shows us how things would look to us if we were as small as a flea, or plankton, or even a cell. Once we can see tiny things, they become more real to us. We can begin to observe them and study how they live.

Of course, some tiny things are not alive, like grains of sand, salt or pollen. For those things, we want to see their structure (how they are put together) and how they interact with other matter. The following piece of science history describes just one instance in which magnification was used to observe and understand a process that could not be seen by our eyes alone.

Science History

Brownian Motion

Figure 2.5 Robert Brown

In 1827, Robert Brown sat in his lab, looking into a microscope. He was a botanist studying pollen grains. Pollen is used by plants to reproduce (have babies). The pollen grains were in a sample of water. Dr. Brown looked closely at them, observing their structure and making notes. He started to notice that the grains were moving. They were jiggling around, sometimes even jumping about in a zigzag motion. What on earth was going on? He wasn't sure, but decided to call this movement Brownian Motion.

Other scientists had also noticed this strange jiggling, zigzag movement. They saw it in all kinds of dust, anytime the dust was observed in a liquid sample. It is important to say here that dust, like pollen, is not alive and should not be moving on its own.

Figure 2.6 Einstein

Other scientists, including Albert Einstein, continued to work on the problem. They finally came up with the cause of Brownian Motion. Here it is: **Water is not just water!**

No, water is actually a whole bunch of water particles that are all mushed together to form the wet liquid. Each of these water particles moves against the ones near it, slipping and sliding around each other. We cannot see them with our eyes, and Robert Brown could not see them with his microscope, but they are there. We know that because we saw how the water molecules interacted with the pollen grains.

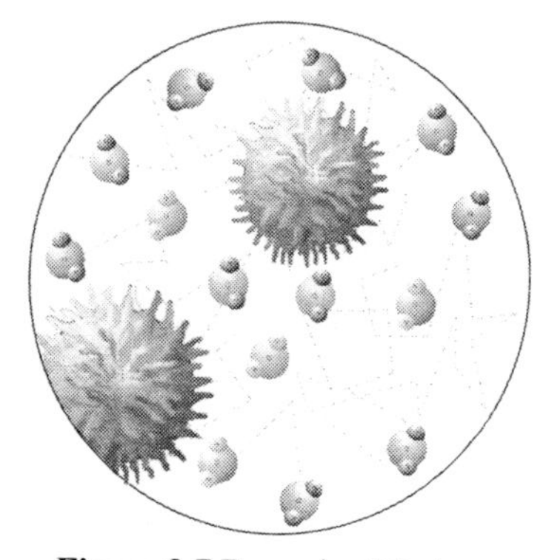
Figure 2.7 Brownian Motion

When pollen grains are placed in the water, the movement of the water particles moves the grains. The motion of the grains actually led to a greater understanding of the existence of even smaller things. Today, we know them as atoms and molecules.

The microscopes that we use today are much more powerful than those used by Dr. Brown. With some types of microscopes, we can actually see those atoms and molecules. These tiny pieces of matter are the building blocks that define all of the matter on Earth. Everything we see and touch is made of atoms and molecules. You will not be surprised to know that an atom is also the sum of its parts. **Atoms** are made of **protons**, **neutrons** and **electrons**. We'll look at these subatomic particles again in Chapter 5.

Activity

The following pictures have been magnified. It is your job to identify them. First, make observations about these images and write them down in the space provided. Next, decide how much you think that they have been magnified: 2x, 10x, 25x, 50x, 100x or more? Write that down. Finally, compare your notes with your fellow scientists. (There is one sitting at the desk next to you!)

dime

dragonfly wing

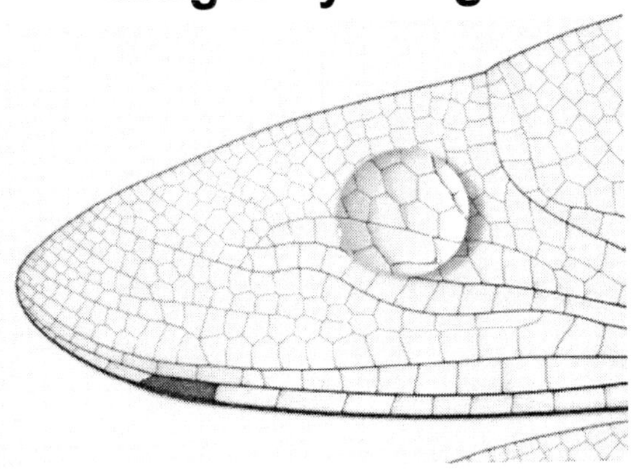

flea

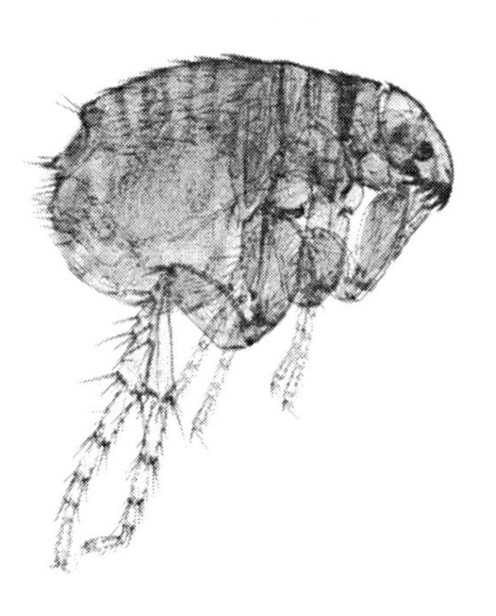

circuit board

crystal of table salt

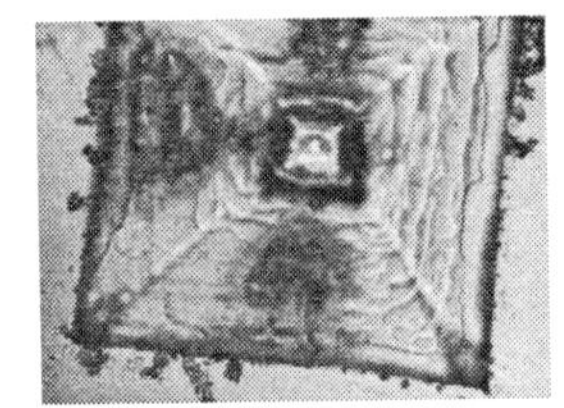

CHAPTER 2 REVIEW

1. What does the objective on a microscope do?
   A. It magnifies.
   B. It focuses.
   C. It observes.
   D. It photographs.

2. Melissa's fingernails are polished a very glossy pink. How do you think they will appear under 100x magnification?
   A. They will be totally smooth.
   B. Pits and ridges will be seen.
   C. They will actually be blue.
   D. You will see through the polish to the nail underneath.

3. An unfocused image is
   A. sharp and clear.
   B. blurry.
   C. patterned.
   D. black.

4. Which of the following objects would you use a magnifying glass to see?
   A. a speck of dust
   B. a television set
   C. a splinter in your toe
   D. your Labrador retriever

5. Which of the following would be hard to see, even if you did use a magnifying glass?
   A. a speck of dust
   B. a television set
   C. a splinter in your toe
   D. you Labrador retriever

Challenge Question

Look back at Figure 2.3 and decide the best way to carry a microscope. Be sure to ask your teacher if you are right, so that you know the correct answer when the time comes. Microscopes are expensive!

Chapter 3
Physical Properties and Changes

Physical Properties of Matter

A **property** describes how matter looks, feels or interacts with other matter. A **physical property** is anything that can change without changing the true identity of the substance that we are observing. Think of dressing up in a Halloween costume. Let's say that you are going to be a dragon. You paint your face with scales if you are "artsy" enough. Make some spines into a crest for your head and back. Create a new, fire-breathing mouth. Wear something dragon-like…you get the picture.

Figure 3.1 Boy and Dragon – Boy

By dressing up as a dragon, you have changed the physical properties that make up your appearance. You have not changed yourself — because we can tell that you are not *really* a dragon.

Matter has many physical properties. In Chapter 1, we described a few of them, including **color, odor and texture**. In Table 3.1, we list a few more.

Table 3.1 Physical Properties of Matter

Physical Property	What it Describes
state (or phase)	solid, liquid, gas
melting point	the temperature at which a solid becomes a liquid
boiling point	the temperature at which a liquid becomes a gas
electrical conductivity	the ability to carry electrical current (electricity)
thermal conductivity	the ability to transfer heat
magnetism	the ability to attract or repulse other matter

Matter on the Move

Physical properties are ways in which we describe and classify matter. **Physical changes** are things that change matter from one form to another. In fact, matter is constantly changing. To follow are two examples. Let's look at each one and see what caused the physical change.

1. *An ice cube melts in your soft drink.* An ice cube is very cold (Figure 3.2A). So cold that the liquid water has frozen into a solid. A soft drink is not as cold as your ice cube, even if the soft drink just came out of the refrigerator. So the ice cube and the soft drink are at different temperatures. The soft drink is warmer than the ice cube. What if you pour the soft drink over the ice cubes (Figure 3.2B)? What is the result of their interaction? The heat from the soft drink moves into the ice cube and melts it. As it melts, the solid ice cubes get smaller (Figure 3.2C). *This is a physical change.* The water in the ice cube does not turn into anything else, like potato chips or marbles. It is still water, but it is in a different physical state. It is now liquid instead of solid (Figure 3.2D).

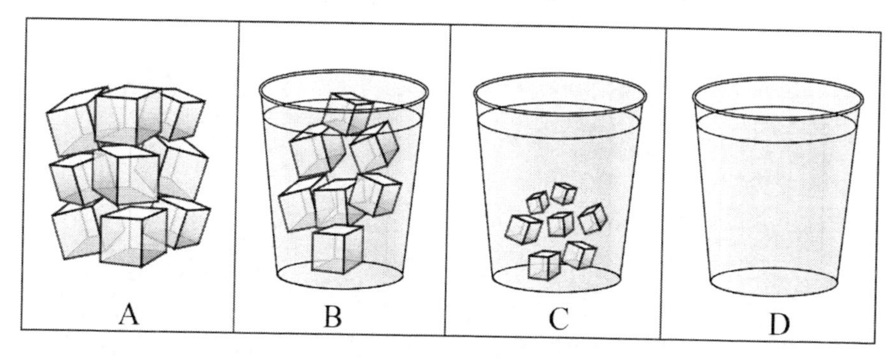

Figure 3.2 Ice Melting in Four Stages

2. *Paper is torn apart.* Paper looks smooth and solid, right? As we see it, it is. But let's look at a piece of paper under the microscope. Under magnification, you will see that it is made of tiny fibers. (The fibers are from the tree wood used to make paper.) These fibers are all wrapped around each other in a disorganized jumble. When you tear a piece of paper, you rip those fibers away from each other. *This is a physical change.* The piece of paper has not changed into two pieces of plastic or metal. The piece has been divided, but it is still paper.

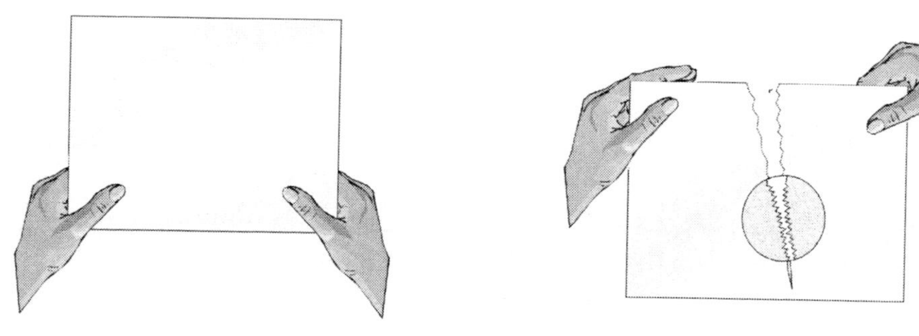

Figure 3.3 Paper Ripping

MIXTURES

When two or more substances combine physically, it is called a **mixture**. What do we mean by "combine physically"? We mean that the identity of each substance does not change when they are mixed.

Think about this. If you drop pebbles into a container of water, what happens? They do not change into something else, do they? No. They each keep their own identity as pebbles and as water. They form a mixture.

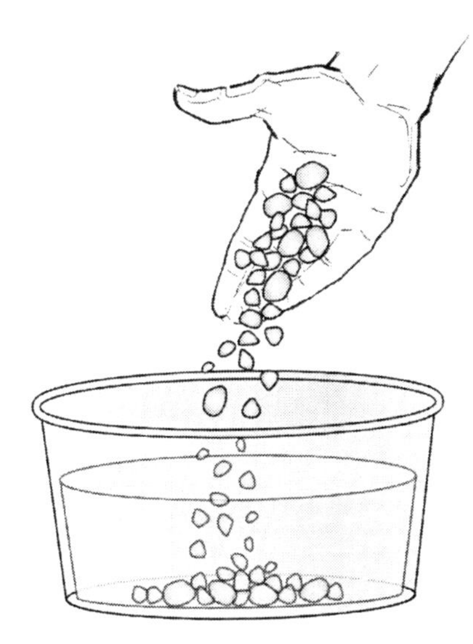

Figure 3.4 A Mixture of Pebbles and Water

Let's mix some other things. How about water and sugar? Put a spoonful of sugar into a glass of water and stir it up. What happens? The sugar ...disappears. When a solid is mixed into a liquid and disappears, we say the solid has **dissolved** (Figure 3.5). The result is a solution. A **solution** is a mixture of one kind of matter (sugar) dissolved in another kind of matter (water). This kind of mixture is the same throughout. As long as it is stirred well, the sugar water will be the same throughout the glass. Scientists call a solution a **homogeneous mixture** (one that is the same throughout).

Figure 3.5 A Homogeneous Mixture

Here is another mixture: put a spoonful of sand into a glass of water. Stir it and stir it and stir it and stir it....OK, stop. If you look in the glass, you will see a spoonful of sand sitting at the bottom of the glass (Figure 3.6). After all that stirring! Just like with the pebbles, the water mixes with the sand, but it does not dissolve the sand. This is not a solution. Scientists call this a **heterogeneous mixture**. A heterogeneous mixture is not the same throughout.

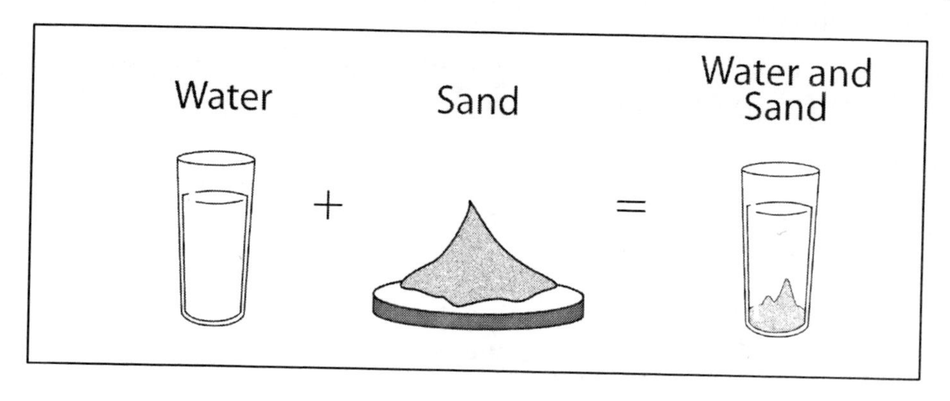

Figure 3.6 A Heterogeneous Mixture

Table 3.2 compares the two kinds of mixture.

Table 3.2 Examples of Mixtures

Homogenous mixtures	Heterogeneous mixture
salt and water	salt and pepper
rubbing alcohol and water	oil and water
Ranch or French salad dressing	Oil and Vinegar salad dressing
chicken broth	vegetable soup

SEPARATING A MIXTURE

A mixture can be separated back into its original substances. It is particularly easy with a heterogeneous mixture. To see this, look back at Figure 3.4. Pour the water off of the pebbles and *TA DA*... they are separated! (You could also just pick the pebbles out of the water). Pouring water off the sand will separate that mixture also.

There might still be some water in your sand, though. In that case, you can filter it. **Filter paper** is a porous paper that will allow tiny water particles (molecules) to pass through, but not larger particles (like pebbles or sand). If you do not have any filter paper, a coffee filter is a good substitute.

Place the filter paper in a funnel, and put it into a container. Pour the mixture into the funnel. The filter paper will stop the sand particles from going through. When water no longer drips from the tip of the funnel, you can remove the filter paper and dry it. Now you have really separated that mixture!

Figure 3.7 Filtering Sand

Now let's separate the sugar water mixture. Can you pour the water off of the sugar to separate the two? No, that won't work.

Haul out the filter paper! Well, actually that will not work either. Filter paper will not separate solutions. The pores of the filter paper are large in comparison to the tiny dissolved sugar molecules. So, the sugar molecules just zip through with the water.

In order to separate a solution we will need to use the process of evaporation. **Evaporation** is what happens when a liquid is turned into a gas. Think of the steam that comes off of a pot of boiling water. That steam is water in the **gas** phase. Before it got hot, it was **liquid** water in the pot.

You do not need to heat your sample to evaporate the water off of it (though that certainly speeds up the process, as you will see in the next chapter.) You can just leave it on the table for a few days. The water will evaporate, drying your solution out the same way that you dried water off the sand in your filter paper.

Whether you heat it or dry it, the result is that the sugar is left behind. Dissolved solids cannot enter the gas phase. They dry back to solids and are left at the bottom of the container.

CHAPTER 3 REVIEW

1. You have just raked the yard. You want to jump in the leaf pile, but you do not want to get stuck with sharp sticks and twigs. What is the best way to separate twigs from leaves?

   A. filtration
   B. evaporation
   C. pick them out
   D. stir them

2. Which of the following is a heterogeneous mixture?

   A. iced tea
   B. lemonade
   C. salad
   D. coffee

3. Melting ice is a physical change. What is freezing water?

   A. a physical change
   B. a physical property
   C. evaporation
   D. a separated mixture

4. Melting and freezing are called "changes of state." What else could you call them?

   A. changes of country
   B. phase changes
   C. physical changes
   D. both B and C

5. Three mixtures are listed below. Which one, if any, could not be separated by evaporation?

   A. salt and water
   B. sawdust and water
   C. iron filings and water
   D. All of these mixtures can be separated by evaporation.

Challenge Question

Dan dumps a bag of ice out onto a hot driveway. Describe what happens to the ice. You get a bonus point for describing how heat is transferred.

Chapter 4
Chemical Properties and Changes

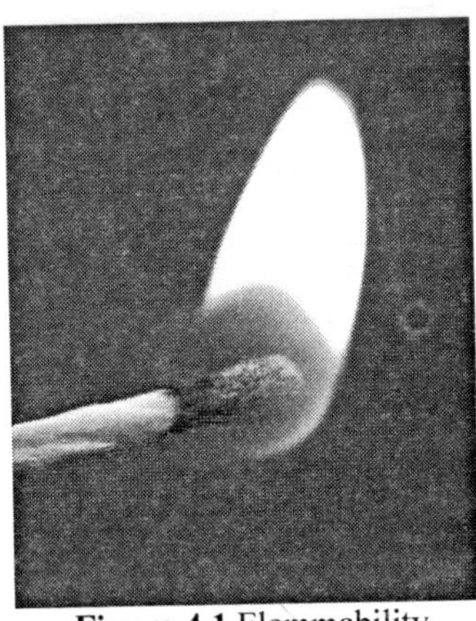
Figure 4.1 Flammability, a Chemical Property

A **chemical property** describes the way a substance may change or react to form a different substance. A **chemical change** happens when one substance changes into another substance. Chemical properties and changes are closely linked. They are also very different from physical properties and changes. For one thing, chemical properties are usually only observed at the time that a chemical change is happening. Here is an example: a piece of paper burns and turns into a black substance. If no one told you that paper could burn, you wouldn't know it – unless it caught fire. So, the chemical property of **flammability** (the ability to burn in the presence of oxygen) is not noticed until the chemical change of **burning** occurs. Table 4.1 contains some important chemical properties.

Table 4.1 Chemical Properties of Matter

Chemical Property	What it Describes
flammability	the ability to burn
toxicity	the degree to which something is poisonous
pH	the acidity of a liquid substance
reactivity	the response of one substance to another
rusting	iron reacting with oxygen

There is another difference. Chemical changes are not usually **reversible**. That means that they cannot be simply changed back to the way they were. Think about the burnt paper. You cannot "unburn" it.

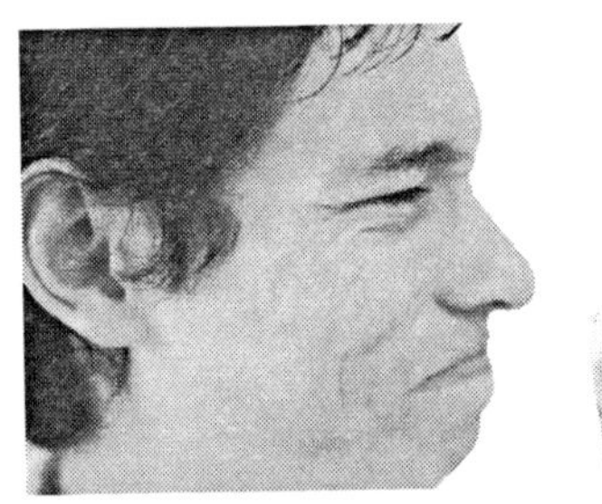

4.2 Sniffing Gross Milk

Some other examples of chemical changes would be cooking a raw egg, milk spoiling or food digesting. Think about whether these changes are reversible or irreversible. For instance, milk needs to be kept in the refrigerator or else it will go sour. When it is left out, the milk gets a sour odor and becomes lumpy. (If you've ever seen or smelled spoiled milk…you'll understand.) So, you can't "unspoil" milk! It is an irreversible chemical change. To identify a chemical change, look for observable signs, as in Table 4.2.

Table 4.2 What to Look For

Possible Signs of a Chemical Change
Heat is produced that was not there before.
Light is produced that was not there before.
Fumes or smoke are produced.
The substances fizz (means a gas phase material is being produced).
Solid substances appear (means a solid phase material is being produced).
The substances change color.
The substances smell different.

Physical and chemical *properties* are ways that scientists classify different types of matter. Physical and chemical *changes* are things that happen to matter to change it from one form to another. As you saw in Chapter 3, matter can be physically combined to form **mixtures**. It can also be chemically combined to form **compounds**. Figure 4.3 should help you to visualize this.

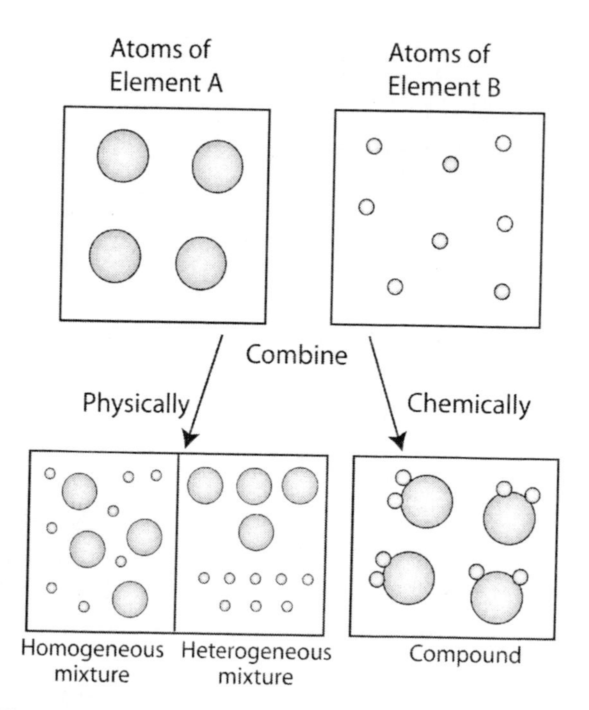

Figure 4.3 Atoms Can Combine Physically and Chemically

Compounds

When atoms of two or more substances combine chemically, they form a **compound.** To "compound" means to combine. That is exactly what a compound is: a combination of the two substances. Look again at Figure 4.3. In a compound, the substances are stuck together, not separated.

This new compound has completely different properties than the individual elements from which it is made. Think about this: hydrogen and oxygen are VERY flammable gases when in their most basic, elemental form. But when we put them together chemically, they form water… which we all know is so stable it is used to *put out* fires. Not only has the chemical property of flammability changed, the physical property of state has also: the two *gases* form a liquid.

Remember that the smallest unit of an element is an atom? In a similar way, the smallest unit of a compound is a **molecule.**

A compound cannot be *physically separated* into its individual components. You cannot, for instance, beat water with a stick to break it into water pieces. It can, however, be *chemically* separated, back into hydrogen and oxygen.

Compounds may also be *physically combined* with other compounds to form mixtures. For instance, table salt is a chemical compound. Put table salt together with sawdust or iron filings and you have a physical mixture of the two substances.

As we will see in the following activity, compounds can also be *chemically separated* or *combined* in processes called **chemical reactions.**

When a chemical reaction takes place, it has to follow the rules. Rules? How can chemical compounds have rules? They do! In fact, everything has to follow this rule. It is a law! **The Law of Conservation of Matter** states that matter cannot be created or destroyed. That means that the mass of any materials that have gone through a chemical change together will be the same after the experiment as it was before the experiment. Read over the following experiment and see this law applied.

A Chemical Change Experiment

You are carrying out a scientific experiment today. You are given your **procedure** (instructions on how to do the experiment).

Step #1: Get your laboratory materials. Use a balance to mass your materials.

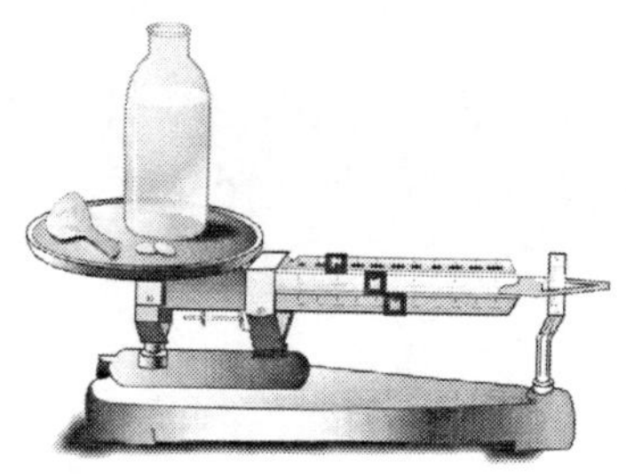

Materials	Mass
Plastic bottle	8 grams
50 milliliters of water	50 grams
1 balloon	2 grams
2 seltzer tablets	10 grams each

Step #2: Record the total mass of your materials: ____________________

Step #3: Put the seltzer tablets inside the balloon and pull the balloon over the mouth of the bottle. Shake the seltzer tablets into the bottle. Draw the experimental set-up in your lab notebook.

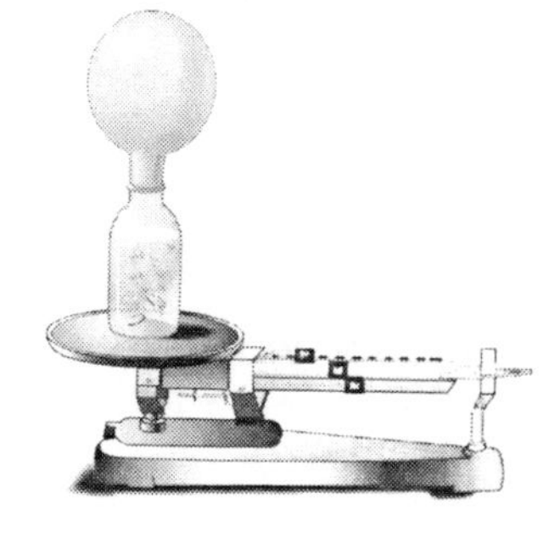

Step #4: Record your observations about the change taking place: ____________________

____

Step #5: Record the mass of the materials again, after the chemical change takes place. Has it changed from the mass that you recorded in Step #2? (Hint: Use the Law of Conservation of Mass to answer this one!) ______________________________

Activity

Describe what is happening in each picture. Will the result be a chemical (C) or physical (P) change? Why?

A

B

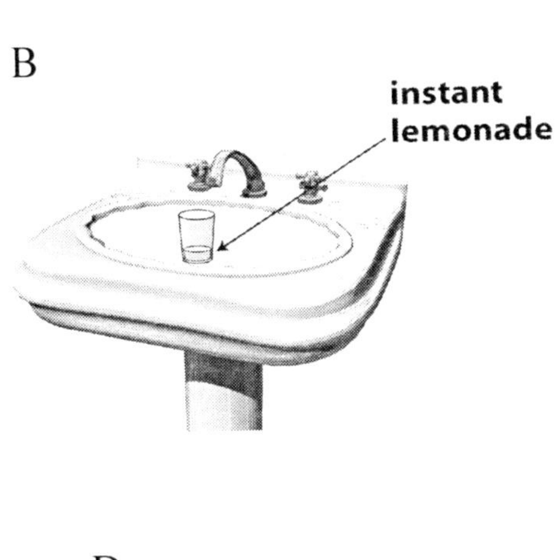

C

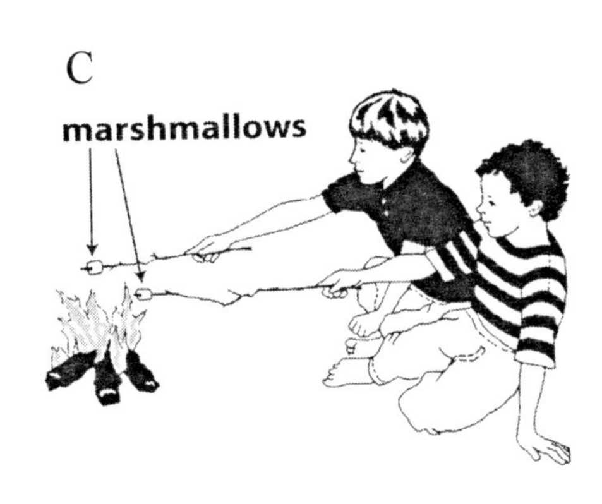

D

E

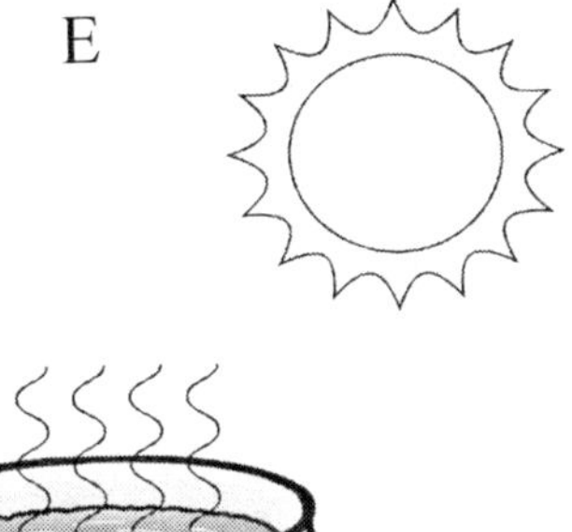

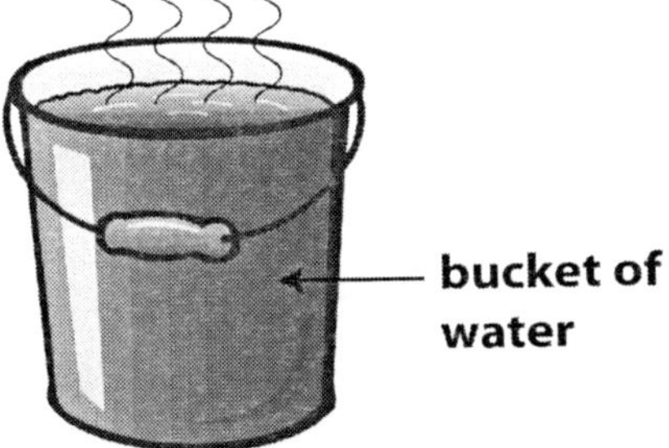

Chapter 4 Review

Use the following chemical change experiment to answer questions 1 – 2.

Tamara records the mass of a whole egg as 35 grams. Then she cracks the egg open. She places the liquid part in a hot pan, and scrambles the egg for breakfast. She records the mass of the egg shell as 2 grams.

1. What is the mass of the egg after Tamara scrambles it?
   A. 33 grams
   B. 35 grams
   C. 37 grams
   D. We cannot tell what the mass is.

2. Which of the following probably alerted Tamara that a chemical change was taking place?
   A. The egg caught fire.
   B. The egg changed to the solid phase.
   C. The egg gave off light.
   D. The stove got cool after the egg was cooked.

3. What is water?
   A. a chemical compound
   B. a physical mixture
   C. a heterogeneous mixture
   D. an atom

4. Which of the following would not always cause a chemical change?
   A. baking
   B. burning
   C. wetting
   D. frying

5. Table salt is mixed with water. Salt water is formed. Which of these materials is a mixture, not a single chemical compound?
   A. water
   B. salt
   C. salt water
   D. All are chemical compounds.

Challenge Question

A fireman uses water to put out a fire. Burning is a chemical change. How do you think the water stops the fire?

Mrs. Magenta Loses Her Marbles

Mrs. Magenta divided her class into groups of ________, each named for a different
(number)

physical property. (Sometimes this worked, and sometimes it didn't. But she was a very

__________ teacher and she liked to try crazy ideas.)
(adjective)

She assigned the groups a box of materials, which they were to sort by __________
(physical

__________. The list of items for each Group was as follows:
property)

1. a __________ ________
   (adjective) (noun)

2. a __________ ________
   (adjective) (noun)

3. a __________ paper plate
   (adjective)

4. a __________ fork
   (adjective)

5. __________ steel cans
   (number)

6. __________ __________ glass marbles
   (number) (adjective)

7. a wooden ________
   (noun)

After all the students got their boxes, the teacher spoke. "Alright, class! Examine the

items in your box. Place all the flammable items in Pile A. Place all your electrically

conducting items in Pile B."

Group __________ __________ sorted their items. Then Group __________
(Adjective) (Physical Property) (Adjective)

__________ __________ over and the two groups spoke. The other Groups
(Physical Property) (verb, past tense)

soon joined them. They stopped sorting and gathered in a ________ on the floor. Mrs.
(shape)

Magenta was sure that this meant they were finished. She examined both of their piles.

Each had organized a Pile C containing item #6, glass marbles.

Mrs. Magenta said "I am so ____________! You all figured it out! Glass is neither
(adjective)

electrically conducting or flammable! This class is so ____________, I could just
(adjective)

____________! ________ bonus points to each of you!"
(action verb) (Number)

Just then, the bell rang, which told the students it was time to ________. ____________
(verb) (Girl's name)

looked at ____________ and said "Wow, Mrs. Magenta was so ____________ with us. It is
(Boy's name) (adjective)

really lucky that Group ____________ ____________ suggested that we play marbles
(Adjective) (Physical Property)

during class!"

Challenge Questions:

1. Which of the items could go into a Pile for electrically insulating items?
2. You defined items #1 and #2, using adjectives and nouns. Could either of those items be sorted into Pile A or Pile B? If not, do they have any other physical or chemical properties that would allow you to define a new Pile?
3. In this MadLib, Groups are named like this: Group ____________ ____________ (Adjective) (Physical Property). Based on this, Alonzo named his Groups "Group Loud Texture", "Group Smooth Smell" and "Group Tiny Toxicity". Which of these names does not include a Physical Property?

Chapter 5
Static Electricity

Here is Jack. He appears to have used a lot of hair gel. Actually, though, Jack is a victim of static electricity. Has this ever happened to you? Why does it happen? Well, we are going to have to look for a moment at what electricity is, and where it comes from, in order to figure that out.

In Chapter 2, we described what atoms were. **Atoms** are the building blocks of other matter. The type of atoms that a piece of matter is made up of define the type of matter that piece is. But even the tiny atom is made up of smaller parts. Atoms are made up of parts called **protons, neutrons** and **electrons**.

Table 5.1 Parts of an Atom

Particle	Location	Charge
Proton	Center of atom	positive
Neutron	Center of atom	neutral
Electron	In a cloud surrounding protons and neutrons	negative

What if we were to tell you that electricity is the result of the movement of one of these subatomic (smaller than an atom) parts? Which one do you think it is? Yes! **Electricity** is the result of moving **electrons**. Different kinds of matter behave differently in the presence of moving electrons. We classify these materials as **electrical conductors** or **electrical insulators**.

Conducting materials allow electricity to flow through them easily. Insulating materials resist the flow of electricity. Table 5.2 contains examples of each.

Table 5.2 Conducting and Insulating Materials

Conductors	Insulators
metals (like copper and aluminum)	rubber, plastics
salt solutions (called electrolytes)	glass
plasmas (the fourth state of matter)	wood

When two different insulating materials come in contact with each other, a surface charge appears. This happens because they are different. Two different materials will have slightly different *degrees* of insulation, even if they are both insulating. For instance, you know that rubber is an insulator. It is a pretty good insulator, in fact. That means that rubber does not allow electricity to flow *easily* through it. But, there is another material called Teflon™ that is an excellent insulator. Teflon™ does not allow electricity to flow through it *at all.*

If you placed a piece of Teflon™ in contact with a piece of rubber, what do you think would happen? Which would become positive and which would become negative? In general, the better insulator will become negative. In this case, that's Teflon™. The reason is that Teflon's electrons can't move. That means that Teflon™ can't carry electricity, as you already know. But it also means that its electrons cannot be taken away. So negative charge builds up on the Teflon™ surface, and positive charge builds up on the rubber surface.

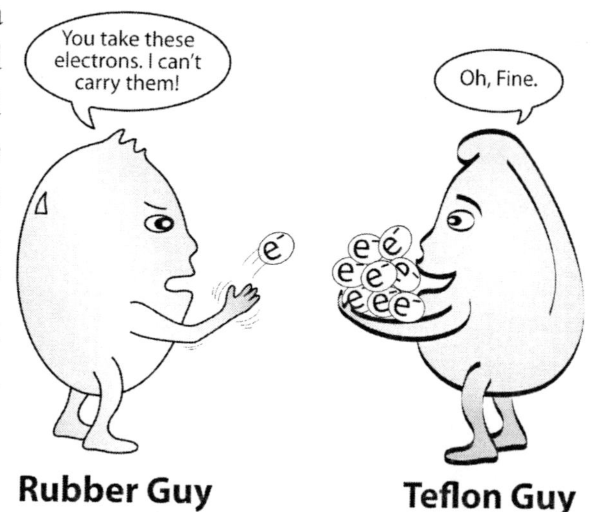

Figure 5.1 Insulators Interacting

We were talking about just placing a piece of rubber in contact with a piece of Teflon™. If we actually rub the two pieces together, the surface charge builds up more quickly.

So, let's go back to Jack. When you take off your hat on a cold winter's day, the hat takes electrons off of your hair. The hat takes on a negative charge and the hairs on your head take on a positive charge. The like charge of the individual hairs causes the hairs to repel one another (push away from each other). So you, like Jack, are standing there with your hair sticking up.

When does it go down? The charge does not just disappear, or slide away. No, it has to be **balanced.** Have you ever heard that opposites attract? Opposite charges are attracted to one another. That means that your positively charged hair is hanging out there trying to attract some negatively charged electrons. When electrons in the air around balance the positive charge, your individual hairs will not repel each other anymore. Your hair will be **neutral** (no charge) and will lie down flat.

Activity 1

Take a balloon and rub it on you hair. Then slowly move the balloon away from your hair. Do you feel any force of attraction as the balloon is attracted to your hair? Write down what has happened to the electrons.

Activity 2

Turn the lights off. Then take your shoes off and shuffle your feet around on the carpet. Then move back over to the light switch and slowly try to touch the metal screw. You might see a blue light flash; this is the electric charge passing between your body and the metal screw. CAREFUL. You will feel a slight shock when the charge disperses.

Science Vocabulary

We are going to examine the phrase **going to ground**. Electricians use it a lot. They say things like

"That charge will go to ground."

"Where's the ground wire?"

Ground is a concept that has to do with charge overflow. If too much charge is delivered too quickly, not even the best conducting material can move the charge. That is why circuits have a ground wire. The "ground" is where excess of charge can be allowed to overflow safely. In some cases, "ground" is actually Earth. A wire in the ground delivers electrons to a huge area (Earth), which will easily balance the excess charge.

CHAPTER 5 REVIEW

1. Which of the following statements best describes static electricity?
   A. Static electricity is the explosion of electrons.
   B. Static electricity builds up when electrons cannot move through a material.
   C. Static electricity comes from the movement of protons.
   D. Static electricity only happens if you rub two insulators together.

2. Which of the following is not an insulator?
   A. silver B. hair C. plastic D. wood

3. Two insulators are placed in contact with each other. What is the only way that no charge will build up?
   A. if they are different materials
   B. if they are the same material
   C. if you separate them with wood
   D. if you keep them perfectly still

4. Which of the following are attracted to each other?
   A. two electrons
   B. two protons
   C. a proton and an electron
   D. a neutron and an electron

5. Which material would be the best choice for an electrical worker to use to protect his hands as he works?
   A. an insulator B. a conductor C. a metal D. water

Challenge Question

In the chapter we talked about the movement of charge when you take your hat off on a winter day. The hat became negatively charged. What happens to that charge if you toss your hat on the counter as you walk in the door?

Chapter 6
Current Electricity

In Chapter 5, we examined static electricity. Static electricity is the build-up of charge between two insulators. In this chapter we will examine current electricity. **Current electricity** is the transfer of electrons through a conducting material. An **electric current** is the flow of charge through a conductor.

Imagine electric current as a line of train cars sitting still on the train tracks. If a moving train car hits the line of resting train cars, the force of the movement will be transferred to each car down the line until the last car rolls away from the line of resting cars. In this example, the train cars are like electrons.

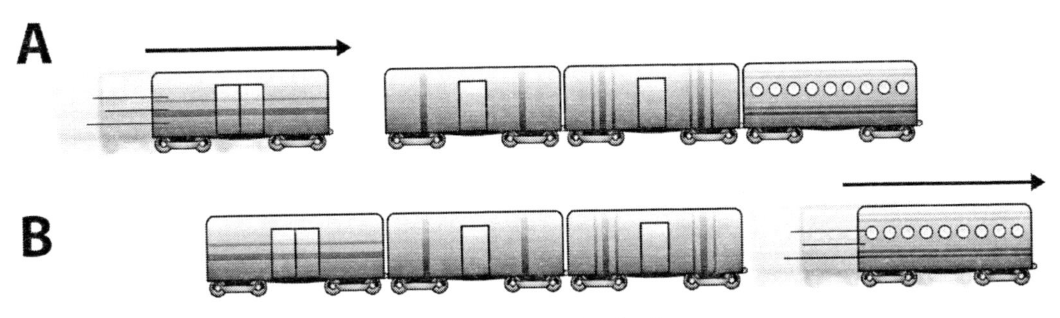

Figure 6.1 A Line of Train Cars

Think of copper atoms sitting next to each other in a copper wire. A moving electron hits the electrons of one copper atom, which in turn hit the electrons of its neighbor, which hit the electrons of its neighbor, … and so on down the line. The neighbors transfer the charge easily and let it go at the end of the wire.

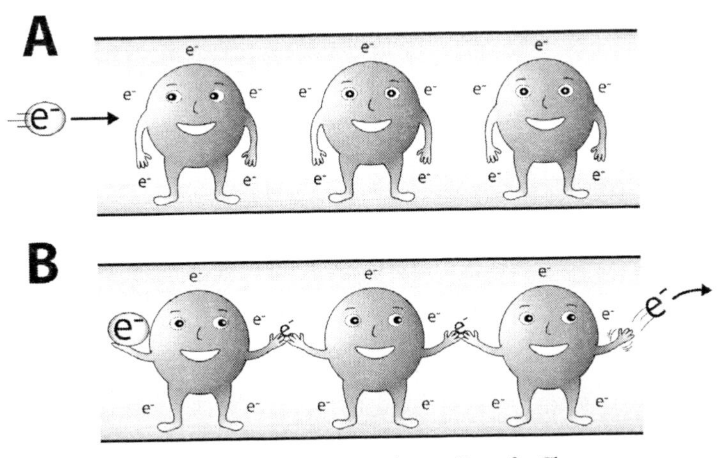

Figure 6.2 A Line of Copper Atoms Transfer Charge

You might be curious about what is at the end of the wire. Where does the electron go once it is released? It goes to a **load**. A load is anything that you want to use electricity to power. A wire, you see, just provides a path for electrons. They stay on the path because the wire is covered in an insulating material (Figure 6.3). Wire allows humans to direct the flow of charge toward a use, like powering a light bulb. The set of paths that we choose to direct that charge through is called an **electrical circuit.**

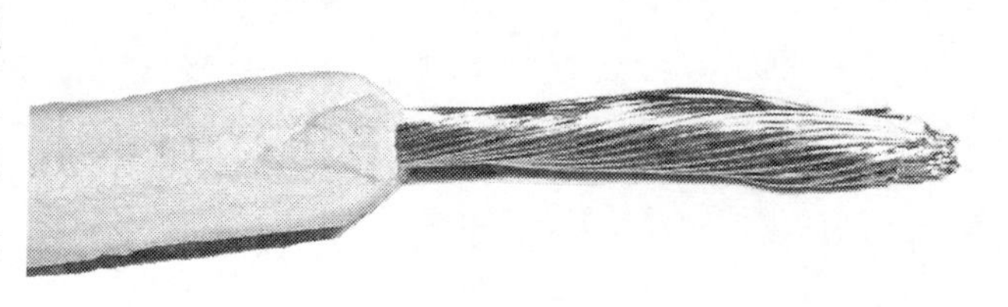

Figure 6.3 Copper Electrical Wire

When we draw electrical circuits, we use the symbols shown in Figure 6.4 to represent voltage sources, resistors and wires. **Batteries** are commonly used as voltage sources. Devices such as radios and televisions draw current from the circuit and so provide resistance to the flow of electricity. These devices, or loads, are usually represented as a simple **resistor** in circuit diagrams.

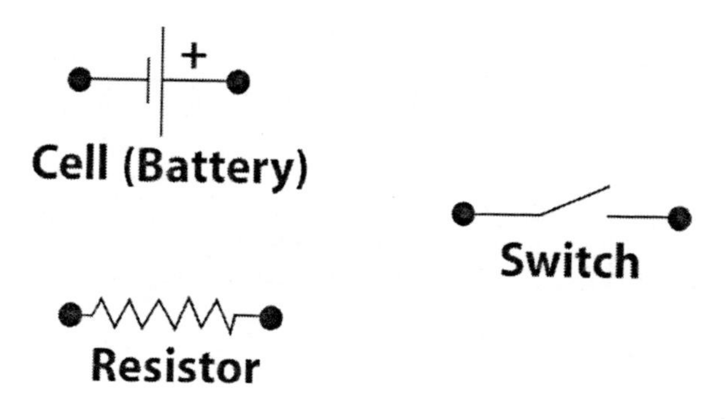

Figure 6.4 Symbols for Circuit Elements

Figure 6.5 shows an electrical circuit, and Figure 6.6 shows the circuit drawing of the same circuit. The three loads (in this case, light bulbs) are powered by a single power source (a battery).

Figure 6.5 Series Circuit

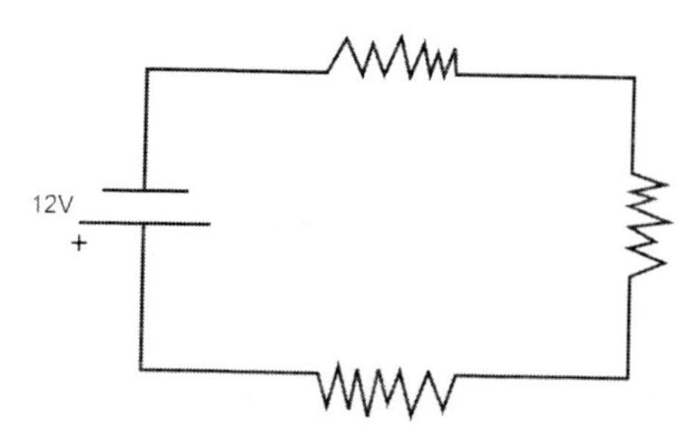

Figure 6.6 Series Circuit

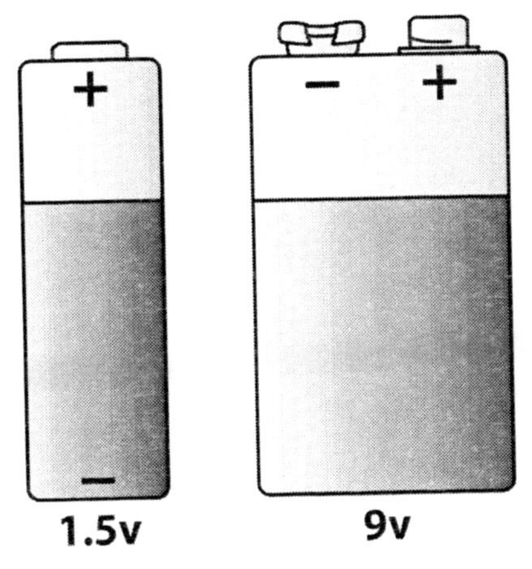

Figure 6.7 Household Batteries

How does this work? The battery is a source of electrical energy. The amount of electrical energy it contains is measured in **volts**. Most common household batteries are 1.5 volt or 9 volt. The chemicals inside the battery allow it to serve as a source of electrical charge.

Once all of the loads are attached, charge flows from the battery through the wire to the first light bulb. There it travels up the filament and across, causing the tungsten that makes up the filament to light up. Charge travels back down to connect to the main circuit. This process occurs in all the light bulbs so quickly that it appears to be all at once. Electrical charge moves very quickly.

The important thing to notice about this circuit is that it is closed. A **closed circuit** allows electricity to flow, and to power the loads attached to it. If a switch in the circuit is opened, electrical flow stops. This is called an **open circuit**. Electricity will not flow in an open circuit.

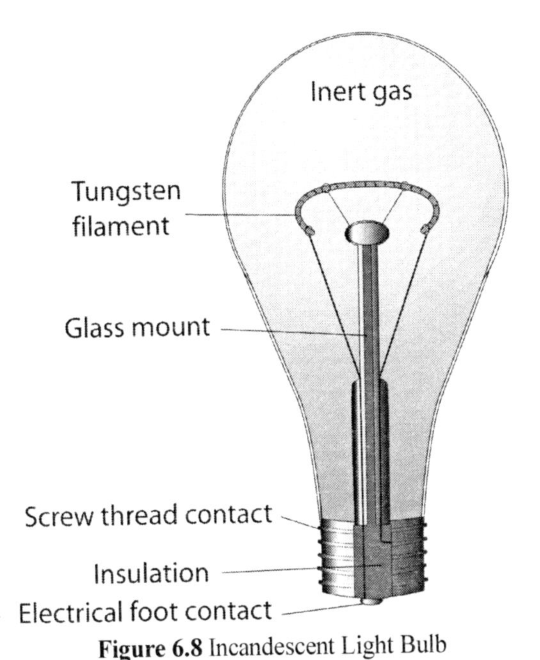

Figure 6.8 Incandescent Light Bulb

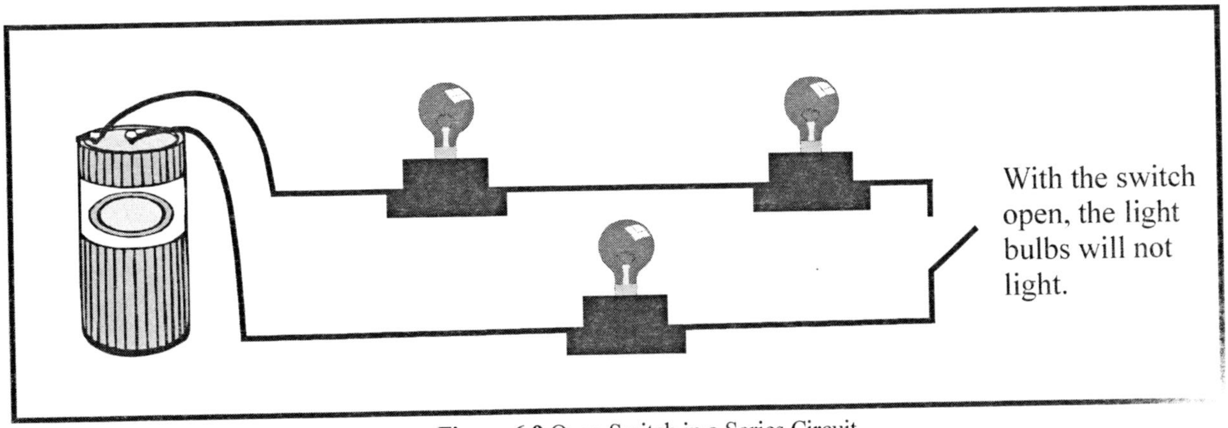

With the switch open, the light bulbs will not light.

Figure 6.9 Open Switch in a Series Circuit

CHAPTER 6 REVIEW

1. A switch is inserted into a series circuit of Christmas lights. During the night, the switch is left open. The lights will

   A. continue to burn.
   B. be turned off.
   C. become a parallel circuit.
   D. burn brighter.

2. Which of the following is a load?

   A. copper wire
   B. a battery
   C. a ceiling fan
   D. a switch

3. What would be the best material to wire an electrical circuit with?

   A. clay
   B. silver
   C. rubber
   D. plastic

4. Which of the following is not powered by electricity?

   A. a television
   B. a microwave
   C. a digital camera
   D. All of these are powered by electricity.

5. When you take the battery out of a flashlight, what have you done?

   A. closed the switch
   B. opened the switch
   C. removed a resistor
   D. both B and C

Challenge Activity

Draw a circuit, using the appropriate circuits symbols, for each of the following scenarios.

1. A battery powers the electrical circuit in a toy ambulance. The ambulance has a speaker (for the siren) and two sets of flashing lights on top. It has an on/off switch.
2. A battery powers a lamp and an electric pencil sharpener on Teri's desk.

Chapter 7
Magnetism

A magnet is defined as anything that produces magnetic field lines. That is a kind of tricky definition, isn't it? Figure 7.1 might help. **Magnetic field lines** are the lines that run from one end of the magnet, outward toward the other end. Magnets have a **north pole** and a **south pole**, and the magnetic field lines run from north to south.

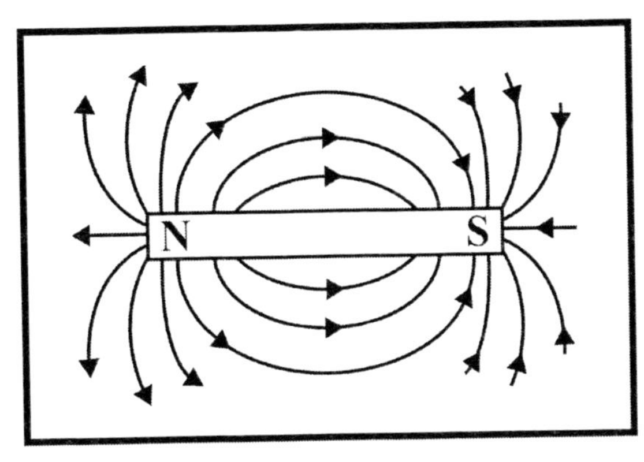

Figure 7.1 Field Lines in a Magnet

A magnet can attract or repel other magnetic objects. As shown Figure 7.2, like poles repel each other and opposites attract. **Repel** means "push away." **Attract** means to "pull closer." So, the north pole of one magnet will repel the north pole of another magnet. That north pole will attract the south pole of another magnet.

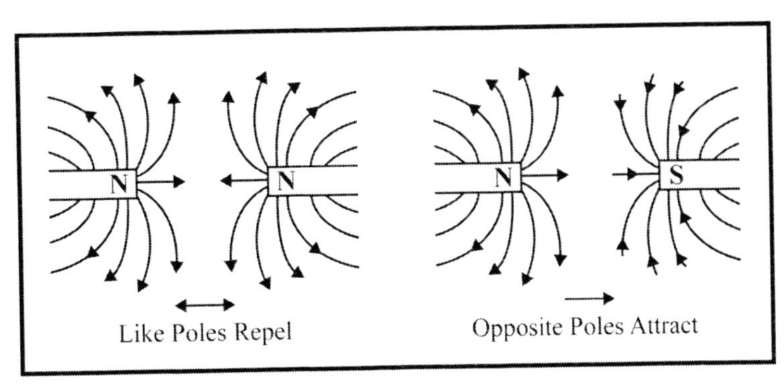

Figure 7.2 Interaction of North and South Poles

Naturally-occurring magnets are made of the mineral **magnetite**. Magnetite is a compound of iron and oxygen. Discovery of this material led to the ancient use of the **lodestone**, a simple compass. In modern times, magnets are usually man-made from a mixture of iron and other metals. A **permanent** magnet is designed to remain magnetic for a long time. One kind of permanent magnet is the refrigerator magnet. Another kind is the bar magnet. A **bar magnet** is a permanent, man-made magnet, shaped in a rectangle. These are usually used to demonstrate magnetism in classrooms.

Let's look inside a bar magnet to find out what makes it magnetic. Using a powerful microscope to look into a magnetic material, you would see that its atoms are aligned in a regular pattern. Each of the particles are arranged with their tiny poles end on end. The series repeats North - South - North - South across the magnet.

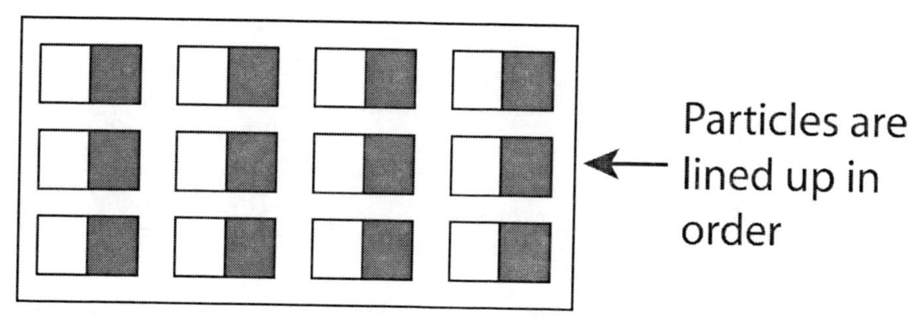

Figure 7.3 A Bar Magnet

So there is no *one* north or south pole inside a magnet, but many. These poles add up to create the total magnetic field. The result is a magnet with an overall north and south pole.

Think of a line of people, each facing the back of the person in front of them. There is no one place where all the faces or the backs are, but the line as a whole has a beginning (the face of the first person) and an end (the back of the last person).

Figure 7.4 A Line of People is Like a Magnet

What happens if you ask the two people in the middle of the line to separate? Now you have two lines, each with a beginning and an end. The same thing happens if you break a magnet in half. You create two new magnets, each with its own north and south pole.

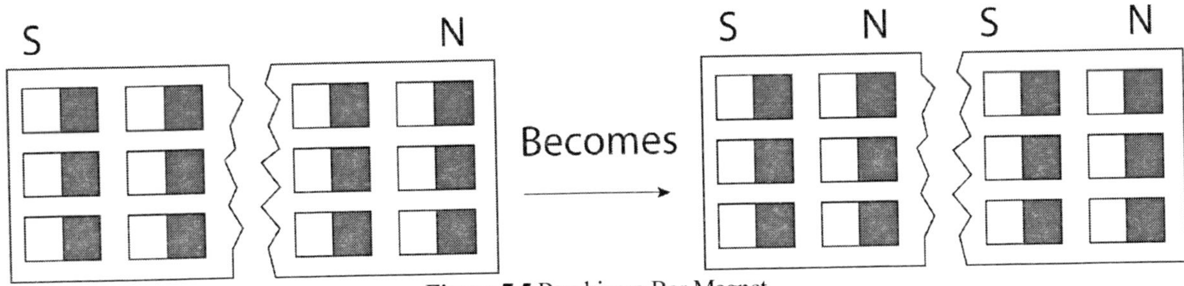

Figure 7.5 Breaking a Bar Magnet

The magnet can be divided again and again, but at some point (depending on its original size), the magnet's particles are no longer arranged in any kind of order. The divided pieces are just too small to keep up a total magnetic field. To see this, think of the line of people again. You can keep dividing the line of people, creating more and smaller lines, until you just have a bunch of individual people standing around.

This is one way to **de-magnetize** a magnet, or remove its magnetic quality. Other ways include heating it to a high temperature or dropping it. Both physical actions will upset the internal order of the magnet and destroy its field.

Electromagnetic Force and Fields

Electromagnets

Electricity and magnetism are actually related forces. One of the fundamental forces in nature is the **electromagnetic force**. The electrical part of this force is responsible for electric charge and the flow of electrons. The magnetic part of the force produces magnetic properties in certain metals.

Since both are parts of the same force, electricity and magnetism are dependent on one another and can influence each other. One example of this interdependence occurs when electrons move or flow. This is an electrical current. The movement of the electrons actually generates a magnetic field. Let's take the simplest example first: let's say that you connect a loop of wire to a battery, as shown in Figure 7.6A. In this arrangement, electrons (e^-) flow through the wire, from the negative (–) terminal of the battery, to the positive (+) terminal of the battery. If you were to view the magnetic field, it would be circling around the outside of the wire, perpendicular to the direction of electron flow. What if you change the direction of electron flow? Well, in order to do that, you turn the battery around, right? The result is a change in direction of the magnetic field (Figure 7.6B).

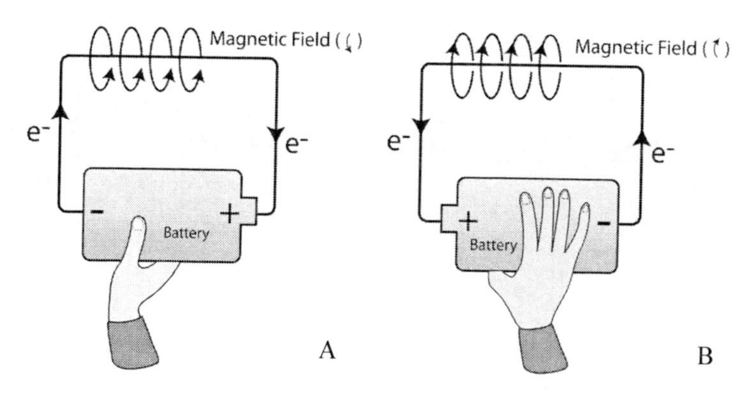

Figure 7.6 Generating a Magnetic Field with Simple Electrical Circuit

We used a word that you may not know: perpendicular. **Perpendicular** means "at a right-angle, or cross-wise, to" something else. Figures 7.7 and 7.8 show the meaning of parallel and perpendicular.

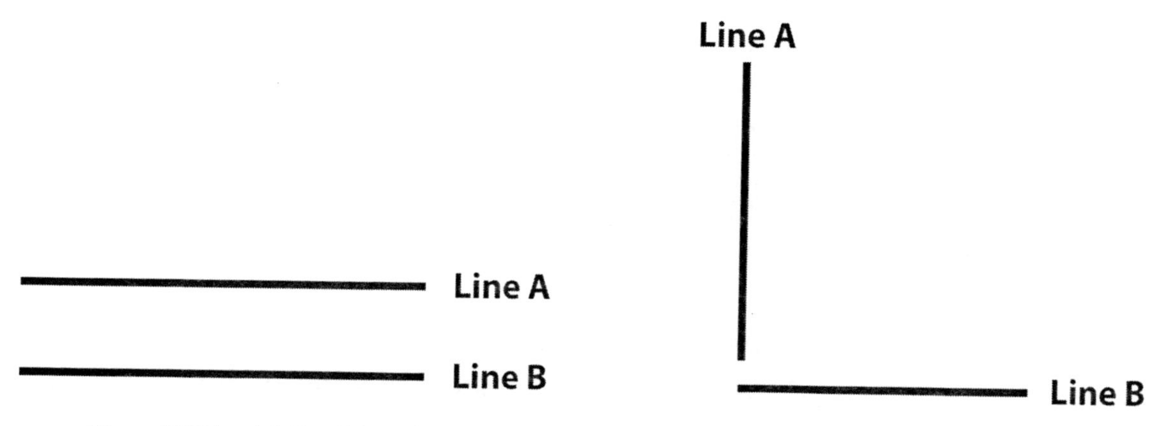

Figure 7.7 Line A is Parallel to Line B

Figure 7.8 Line A is Perpendicular to Line B

Perpendicular is an important word when talking about electricity and magnetism. Remember how we said both forces, electrical and magnetic, were part of the same force, the electromagnetic? Well, the word perpendicular describes how the forces line up. They are always perpendicular to each other.

Now let's go back to our electromagnet. What would make the magnetic field stronger? Coil the wire! When you do this, the magnetic field at every place in the wire will add to the field of the place next to it, because the magnetic field will be going in the same direction in both places. The more coils that you create, the stronger the field becomes.

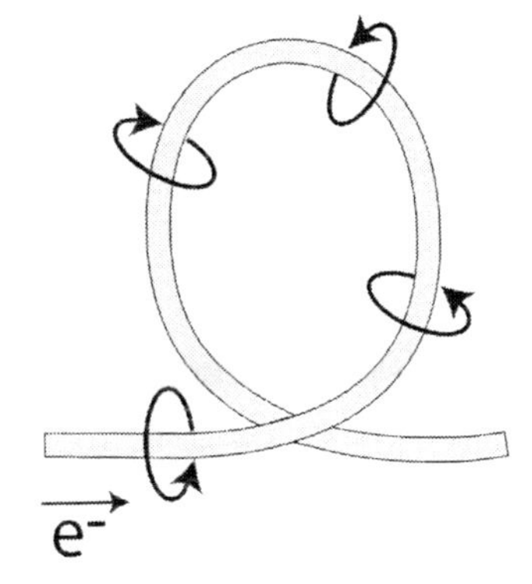

Figure 7.9 Coiling the Wire

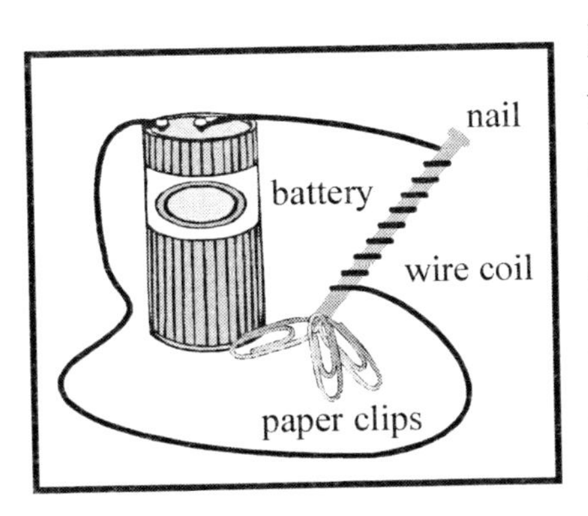

Figure 7.10 Electromagnet

But why is this useful? What can you do with a magnetic field? You can make an electromagnet! As the name implies, an **electromagnet** is a device that becomes magnetic when electricity flows through it.

Let's take the basic set-up from Figure 7.6, add coils to the wire as in Figure 7.10 and, finally, add the *pièce de résistance*...an iron nail! What will happen? Each individual magnetic field adds together to create an overall magnetic field that is quite strong. This turns your everyday nail into a magnet, with north and south poles. This is shown in Figure 7.11. See how the overall magnetic field is perpendicular to the direction the electrons are moving?

You might be wondering what the purpose of the nail is. The nail actually does two things. First, iron is a **ferromagnetic** material. That means it is easy to magnetize. The magnetic field from the coil causes the electrons in the nail's iron (Fe) atoms to line up with the field. If we removed the nail, and the core of the coil was just air, a magnetic field would still be generated around the wire, but it would not be as strong. The magnetic field of the nail adds to the magnetic field in the wire to make a strong overall magnetic field.

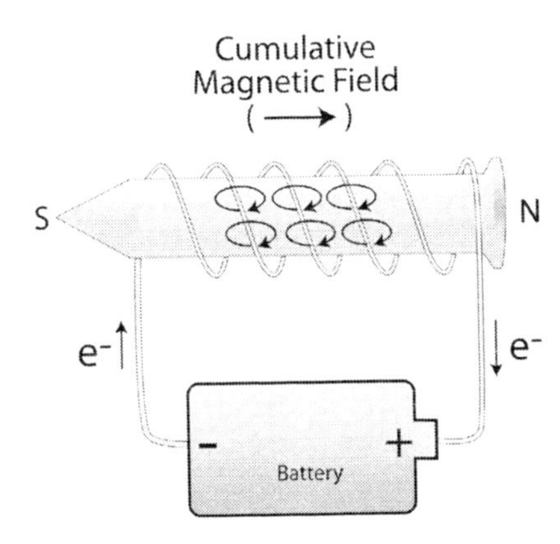

Figure 7.11 A Simple Electromagnet

The second purpose of the nail is to serve as a solid cylinder to wrap the wire around. This keeps the wire coiling in a uniform direction, so that the magnetic fields are aligned properly.

The magnetic field will last as long as the battery continues to deliver electrons (a current). This means that an electromagnet will only work when the circuit is closed and electrons can flow. Also, when the battery "dies," no more magnetic field can be generated.

Figure 7.12 shows our electromagnet without its battery attachments or its nail. This is just a schematic to show you the full effect of the magnetic field (remember, no current, no field!). The field is strongest close to the coils, and gets weaker as you move farther away.

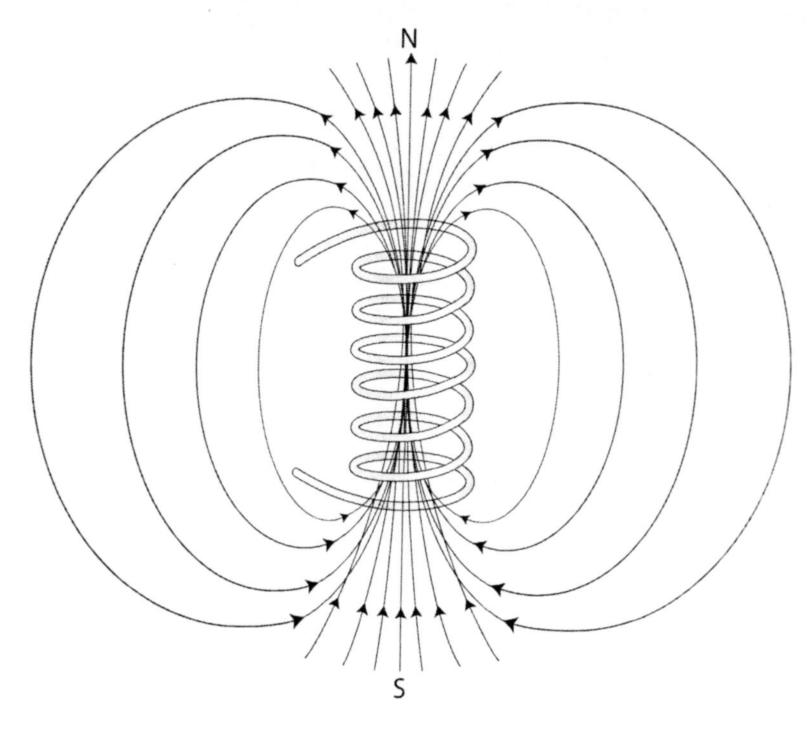

Figure 7.12 The Magnetic Field of an Electromagnet

Electromagnets vs. Bar Magnets

So, how is an electromagnet different from a bar magnet? As you know, a bar magnet is a magnet that produces a permanent magnetic field *naturally*. In other words, current is not applied to the bar magnet to generate the magnetism. Instead, each atom of the magnet is lined up in such a way that their magnetic fields add up to one large field. This means that every atom is a tiny magnet by itself. Lining these atoms up is similar to lining up a bunch of bar magnets.

On the other hand, an electromagnet produces a "temporary" magnetic field from the flow of electrons in the wire. There is no permanent change in the atoms of the wire for them to "hold" the magnetism. So, the magnetic field disappears when the current stops. This can be useful for certain jobs.

One example is the giant electromagnets used to move junk cars. A crane places the electromagnet over a car. The electromagnet is turned on, attracting the metal in the car.

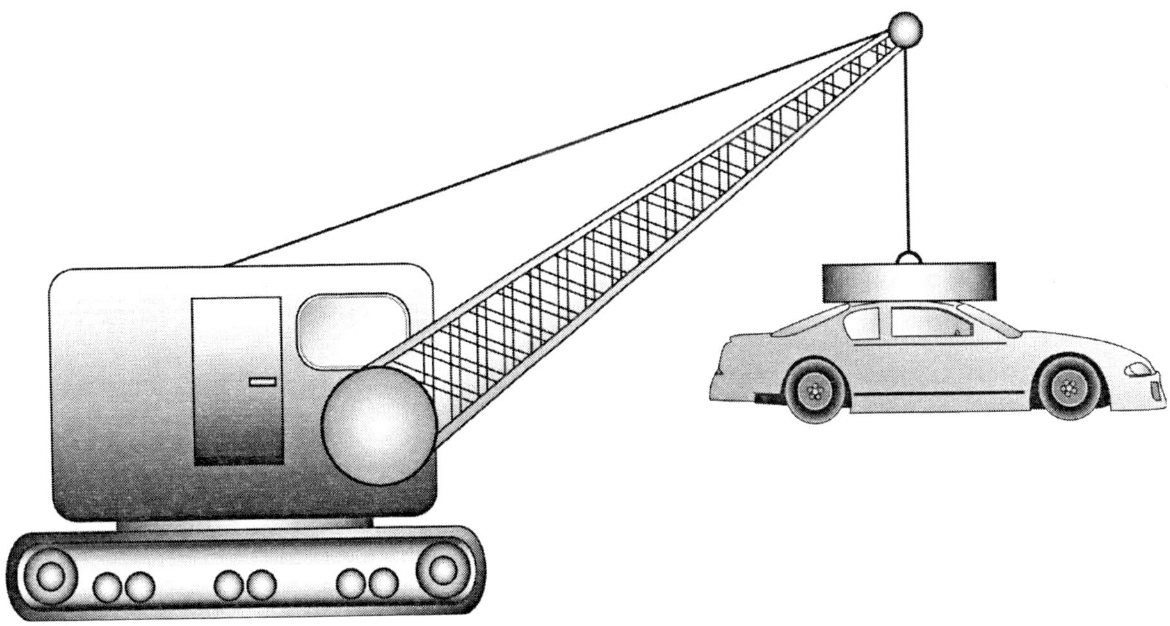

Figure 7.13 A Junkyard Electromagnet

The car is moved to the crusher machine. There, the electromagnet is turned off. With the electrical current gone, the magnetic field disappears and the car drops into the crusher.

A few basic design features of electromagnets should be fairly easy to guess, if you think about it. One way, for instance, to control the strength of the magnetic field would be to increase or decrease the current through the wire. Also, one can increase the magnetic field strength by increasing the number of coils wrapping around the nail.

Very large magnetic fields can be generated by designing the electromagnet correctly, much larger than would be possible by using bar magnets. In 2006, the world's largest electromagnet was built in Switzerland with 20,000 amps flowing through it – that's the same amount of current in a lightning bolt!

CHAPTER 7 REVIEW

1. Which of the following cannot be a magnetic material?

   E. a lodestone
   F. magnetite
   G. air
   H. an iron nail

2. What is the main difference between a bar magnet and an electromagnet?

   A. A bar magnet is not powered by an electrical current.
   B. A bar magnet has no electromagnetic force.
   C. A bar magnet cannot be de-magnetized.
   D. An electromagnet cannot be de-magnetized.

3. Jason places a bar magnet on a flat surface and covers it with a sheet of paper. Then he evenly sprinkles a layer of iron filings on top of the paper. Which of the following diagrams indicates the most likely arrangement of the filings on the paper?

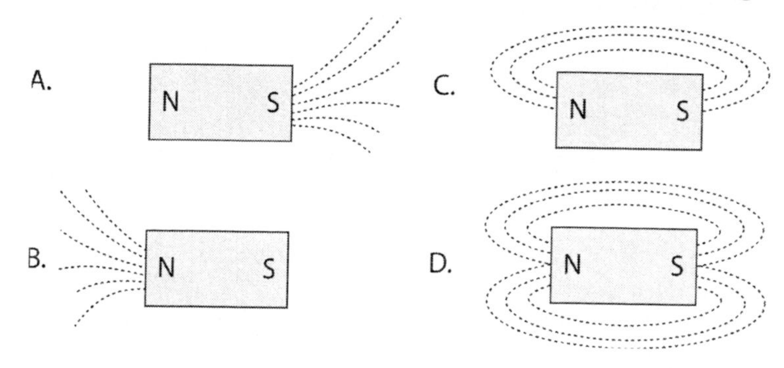

4. Which of the following will attract one another?

   A. the north pole of a magnet and the south pole of another magnet
   B. the north pole of a magnet and the north pole of another magnet
   C. the south pole of a magnet and the south pole of another magnet
   D. all of the above

Challenge Question

How would you change the direction of the magnetic field in a simple electromagnet (like the one shown in Figure 7.10)?

Physical Science
Domain 1 Review

1. Which of the following is not a physical change?
   A. melting point
   B. color
   C. odor
   D. flammability

2. Brownian motion causes
   A. random movement of nonliving particles.
   B. nonliving particles to come to life.
   C. liquid water to turn into nonliving particles.
   D. liquid water to become solid water.

3. A bag of ice is dumped on your concrete driveway on a hot August day. What happens?
   A. The ice undergoes a physical change.
   B. The ice undergoes a chemical change.
   C. The concrete melts.
   D. The ice become concrete.

4. What must be moving for an electrical current to flow?
   A. water molecules
   B. metal atoms
   C. electrons
   D. magnetic particles

5. An old log cabin catches fire. What happens?
   A. The wood undergoes a chemical change.
   B. The wood undergoes a physical change.
   C. The wood turns into air.
   D. The wood turns into dirt.

6. What tool should you use to clearly see the hairs on the legs of a cockroach?
   A. a magnifying glass
   B. an electromagnet
   C. a microscope
   D. a telescope

7. What is the smallest sample of water a scientist can see with a powerful microscope?
   A. a teaspoon
   B. a pint
   C. a milliliter
   D. a molecule

8. What part of a microscope should you adjust to focus the view clearly?
   A. the adjustment knobs
   B. the nosepiece
   C. the arm
   D. the objectives

9. Mr. Rummage runs a salvage company. At the junkyard, he has a large electromagnet. Which of the following items can he pick up with the electromagnet?
   A. old shoes
   B. broken windows
   C. used washing machines
   D. recycled cardboard

10. Wanda has built an electrical circuit. She turns it on and watches a light bulb burn brightly. When class is over, the teacher asks that everyone take apart his or her circuits. What should Wanda do first?
    A. Open the switch to turn off the circuit.
    B. Close the switch to turn off the circuit.
    C. De-magnetize the circuit.
    D. Add another resistor to the circuit.

Domain 2
Chapters 8 – 13

Chapter 8: Classification

S5L1: Students will classify organisms into groups and relate how they determined the groups with how and why scientists use classification.

Chapter 9: Animal Groups

S5L1: Students will classify organisms into groups and relate how they determined the groups with how and why scientists use classification.

a. Demonstrate how animals are sorted into groups (vertebrate and invertebrate) and how vertebrates are sorted into groups (fish, amphibian, reptile, bird and mammal).

Chapter 10: Plant Groups

S5L1: Students will classify organisms into groups and relate how they determined the groups with how and why scientists use classification.

b. Demonstrate how plants are sorted into groups.

Chapter 11: Parts of a Cell

S5L3: Students will diagram and label parts of various cells (plant, animal, single-celled, multi-celled).

a. Use magnifiers such as microscopes or hand lenses to observe cells and their structure.

b. Identify parts of a plant cell (membrane, wall, cytoplasm, nucleus, chloroplasts) and of an animal (membrane, cytoplasm and nucleus) and determine the function of the parts.

c. Explain how cells in multi-celled organisms are similar and different in structure and function to single-celled organisms.

Chapter 12: I Like your Genes

S5L2: Students will recognize that offspring can resemble parents in inherited traits and learned behaviors.

a. Compare and contrast the characteristics of learning behaviors and of inherited traits.

b. Discuss what a gene is and the role genes play in the transfer of traits.

Teacher note: Be sensitive to this topic since biological parents may be unavailable.

Chapter 13 Surrounded by Germs

S5L4: Students will relate how microorganisms benefit or harm larger organisms.

a. Identify beneficial microorganisms and explain why they are beneficial.

b. Identify harmful microorganisms and explain why they are harmful.

Chapter 8
Classification

It is a hot summer day. You are sitting in your front yard, watching the sprinkler water the garden. Every time the sprinkler spins around you see a tiny little rainbow appear. Soon you begin to think that being at home is the hottest, most boring thing ever. You find yourself remembering your last vacation. Last week you were in a very different place — you were visiting the beach.

Figure 8.1 Beach

Figure 8.2 Clam

At the beach, everything is so much more exciting. Even the air smells different! You remember walking on the beach each morning and seeing lots of new ocean creatures washed up by the tide. There were interesting brown plant-like things, small-shelled animal-like things and even some things that didn't look like plants or animals. As you think about all the new creatures you saw on your vacation, you begin to think of one in particular; the small-shelled creatures living in the sand.

Figure 8.3 Quahogs

Your parents called these animals "clams" but your grandfather said they were called "quahogs." Your best friend's mom said they are called "southern pigtoes." Who is right? This problem begins to trouble you as you start thinking: maybe your grandfather was confused about the living thing you described to him. Maybe your parents were wrong. Maybe your best friend's mom really is crazy. You begin to wonder, why is there so much doubt about the name of this living thing? How do you know who is right and who is wrong?

As it turns out, scientists have devised a method to solve these kinds of problems. Scientists use a process called **classification** to group things together based on shared characteristics. When scientists group living things together, it is called **biological classification**, or **taxonomy**.

Scientists start by classifying living things, also called organisms, into one of six kingdoms. A kingdom is the largest group of living things and contains the most organisms. The six kingdoms are Eubacteria, Archaebacteria, Protista, Fungi, Plantae and Animalia. Kingdoms divide into smaller and smaller groups, each containing fewer and fewer organisms. You will learn the names of these groups later in school. The last two groups of living things are the specific genus and species. A species is the most exact category and contains the fewest organisms. A **species** is a group of organisms that can breed in the wild and produce fertile offspring (children).

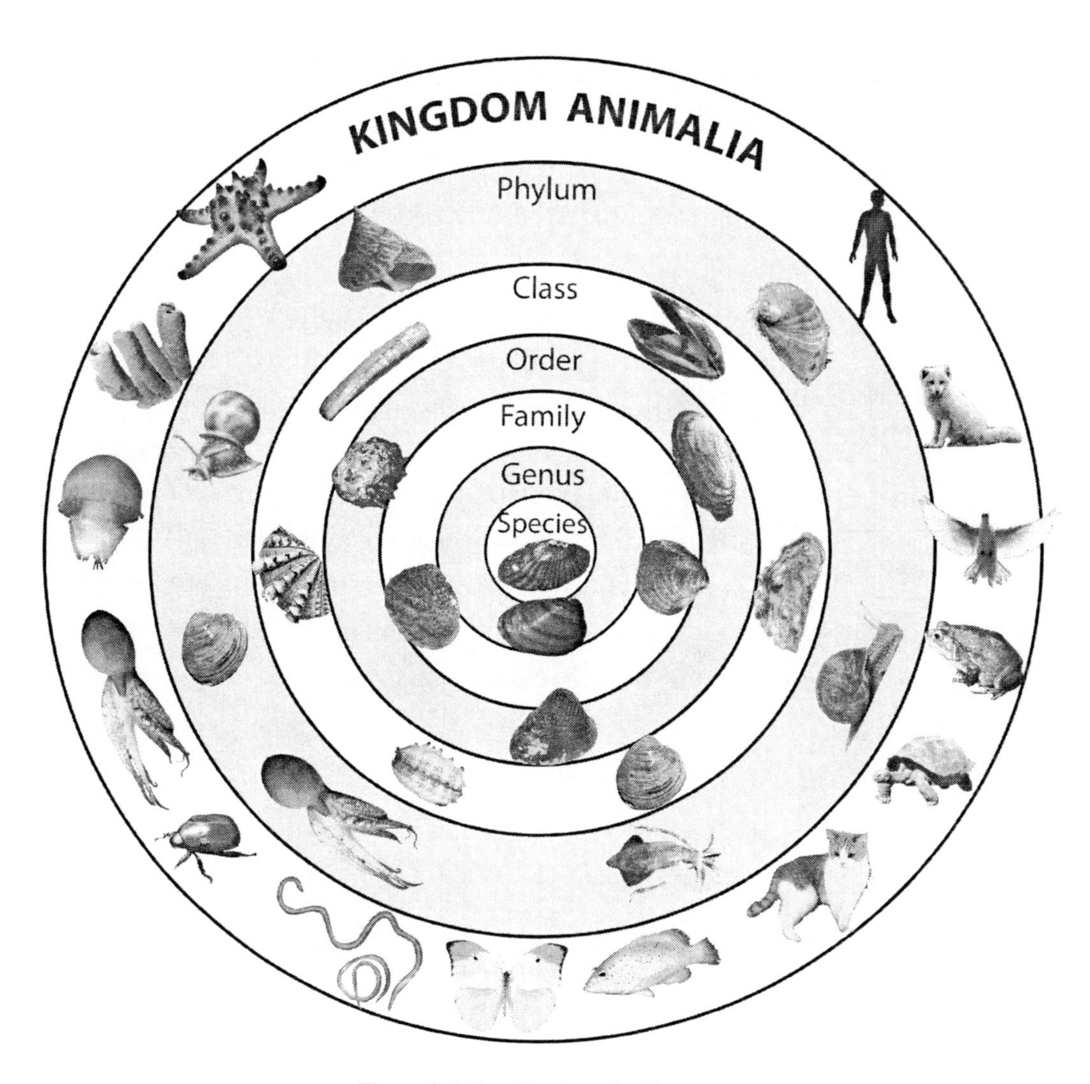

Figure 8.4 Classification of a Clam

Examine Figure 8.4 again. Did you notice how the animals in the model got more similar the closer they were to the center? This model shows how each classification level becomes more specific in each successive group. There is a wider variety of shells types in a class than in a genus. Clams classified in the same species are the most alike and clams classified in the same class can be very diverse.

Figure 8.5 *Pleurobema georgianum*

When scientists classify a living thing, they give it a **scientific name**. A scientific name is a Latin name assigned to an organism which represents the genus and species given to the living thing. In this way, scientists clearly identify living things and stop any confusion caused by common names. (Remember clams, quahogs and southern pigtoes?) Now, if someone told you the living thing you were interested in was called *Pleurobema georgianum* you would know exactly what they were talking about (or at least you could look it up!). Through classification, scientists make it easier to study the millions of different living things found on Earth.

You might be wondering: how do scientists group living things? How do they know which group each organism belongs in? Well, scientists use several methods to classify living things. One method is by looking at the appearance of the organism. Other qualities used to classify organisms are the organism's DNA or the life cycle of the organism. In the next chapter, we will get some practice classifying living things into different groups.

Activity

You teacher has placed tubs full of different kinds of sea shells in the center of your table. Select one shell from the tub. Carefully examine your shell and record characteristics of your shell in your notebook with words or drawings. Then place your shell back into the tub and carefully mix the tub contents. Try to find your shell based on your description. Did you get the correct shell on your first try? How can you improve your description? Try the activity again with a different shell.

CHAPTER 8 REVIEW

1. How do scientists NOT group living things?
   A. by their physical structure
   B. by their birthday
   C. by their DNA
   D. by their lifecycle

2. Why is a scientific name better than a common name?
   A. it keeps people from being confused
   B. it confuses people
   C. it helps organisms know where they belong
   D. it helps people know what is alive

3. What is the smallest group of living things called?
   A. kingdom B. plants C. animals D. species

4. What is the main difference between animals in the same genus and animals in the same phylum?
   A. Animals in the same phylum are very similar and animals in the same genus are very different.
   B. Animals in the same phylum are very different and animals in the same genus are very similar.
   C. Animals in the same phylum are exactly the same and animals in the same genus are very different.
   D. Animals in the same phylum are all the same color while animals in the genus are different colors.

5. What is grouping living things known as?
   A. taxonomy
   B. groupification
   C. species
   D. clams

Challenge Question

Develop your own method for classifying things found at the beach. Write down instructions for how to use your method of classification and give it to your lab partner. Then have them use your new method to classify things found at the beach.

Chapter 9
Animal Groups

Figure 9.1 At the Fishing Pier

Now that you know what all those shelled creatures living in the sand are called, you begin thinking about all the other types of animals you observed on your trip. You remember visiting the fishing pier where people were fishing for all different types of animals. You remember seeing the crab fisherman pull up their cages full of blue crabs and the oyster fisherman hauling in buckets of oysters. Then there is the sheer variety of fish that were caught: flounder, mackerel, amberjack, snook... and how about that small hammerhead shark! Could you ever forget seeing dolphins hunt for fish under the pier? Or the seagulls and pelicans swooping overhead?

Remembering that day makes you think about the diversity of animals on Earth. How many different kinds of animals are there? Start thinking about all the different kinds of animals you know. Quickly, jot down all the different animals you can think of. You can use the space below.

Remember that one of the six kingdoms mentioned in the previous chapter was Animalia. Well, as it turns out, this kingdom is the animal kingdom. As you know, there are many different kinds of animals. How is one type of animal separated from another? In fact, what makes animals different from plants? To answer these questions, let's look at what all living things have in common and narrow it down from there.

All living things are made up of cells. Animals share three main characteristics. First, animals are made of many different types of cells and are called multicellular. **Multicellular** means made of many cells. Second, all animals are able to move. Animals are able to move around in the environment during some point in their life. Lastly, animals cannot make their own food. Animals eat other organisms for food. Animals can eat plants, other animals or dead organisms for food.

Table 9.1 List of Characteristics

Animals
are made of many cells
move
eat other things

Let's look back at the animals you wrote down. No doubt there is a wide variety of organisms listed. Now that you know what makes an animal an animal, see if you can add any more animals to your list.

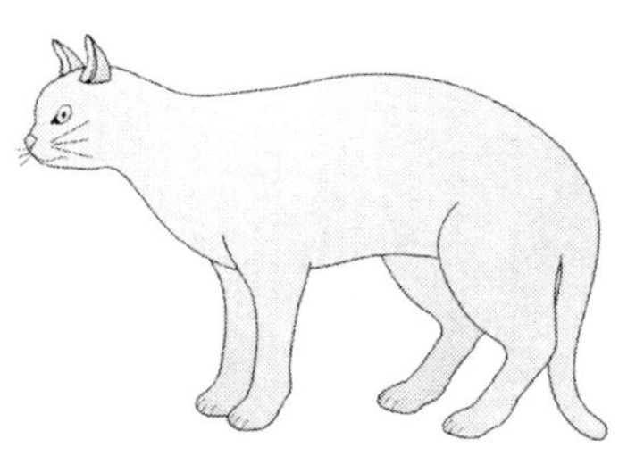

Figure 9.2 Vertebrate

Recall from the previous chapter, classifying is placing things into smaller and smaller groups. Try to think of a way to classify the animals you listed into two distinct groups. Remember, animals are first grouped based on physical structure.

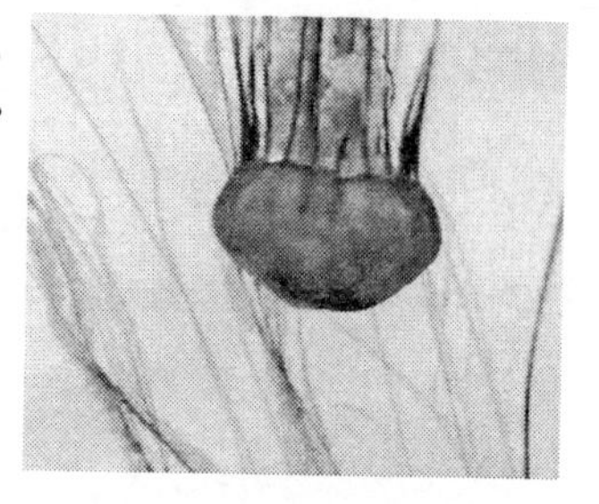

Figure 9.3 Invertebrate

In biology, the two main groups of animals are vertebrates and invertebrates. A **vertebrate** is an animal with a backbone, or spine. Some examples of vertebrates are *humans*, *dogs*, *cats*, *goldfish* or *birds*. An **invertebrate** is an animal that does not have a backbone. Examples of invertebrates include *worms*, *insects*, *clams*, *starfish*, *sponges* or *jellyfish*.

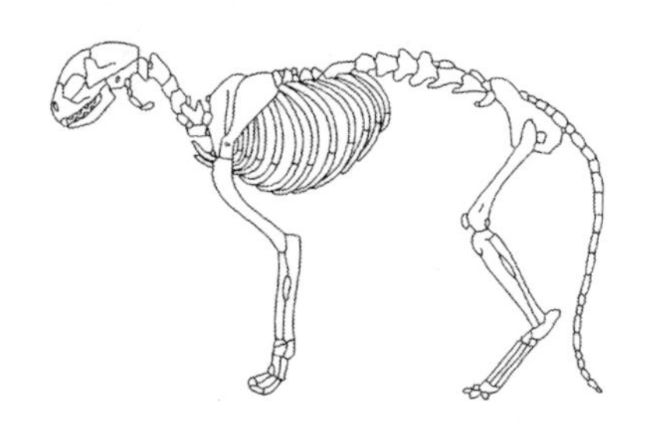

Figure 9.4 Internal Skeleton

Examine your list and pick out all the vertebrates listed. What physical traits do they all share? First of all, they all have an **internal skeleton**, or a skeleton located on the inside of their body.

Inquiry Challenge

Examine the pictures of animals provided by your teacher. Decide if each animal is a vertebrate or invertebrate. Of the vertebrate animals, decide how to group the animals into five different groups. What traits did you use to group the vertebrates?

Vertebrates are also classified according to their physical structure. While doing the activity above you should have placed animals that look similar in the same group. The five main groups of vertebrates are: *fish*, *amphibians*, *reptiles*, *birds* and *mammals*. Look at your group. Do they resemble the five main groups of vertebrates listed? Major characteristics of each vertebrate group are described below.

Figure 9.5 Example of a Bony Fish

Figure 9.6 Shark

Fish are vertebrates that live in the water. Examples of fish include *tuna*, *bass*, *sharks*, *rays* and *hagfish*. Fish have **gills** that they use to breathe. Most fish have **scales**, or hard bony structures that protect the fish. Most fish also have fins that help balance and move the fish through the water. A **fin** is a broad, flat structure used for movement. Some fish have a skeleton made of bone, very similar to your bony skeleton. Other fish, like sharks, have a skeleton made of cartilage. **Cartilage** is a tissue that is tough but also flexible, like the tip of your nose. With a few exceptions fish are **exothermic**, meaning their body temperature changes with the environment. Fish can be found in the ocean, rivers, lakes, swamps and marshes.

Amphibians are vertebrates that can live on land or in the water. Some examples of amphibians are frogs, toads, salamanders and newts. During their lives, amphibians change from an immature tadpole into an adult. Figure 9.7 shows the frog life cycle. Notice, young amphibians look very different from the adult animals. The eggs of amphibians are very similar to fish eggs. The eggs have a thin skin and no shell. As a result, the eggs must remain underwater or moist while the young animal is growing inside. Because of the

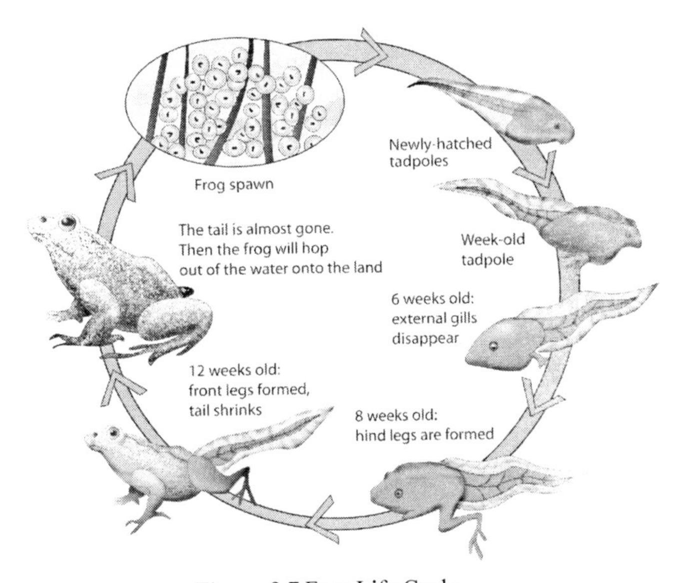

Figure 9.7 Frog Life Cycle

egg structure, amphibians must remain near water to reproduce. When they are young, amphibians use gills to breathe underwater. But when they get older, they use lungs to breathe air. Amphibians have thin **sensitive skin** they use to breathe. Amphibians have a bony internal skeleton and are exothermic. Amphibians can live in a variety of environments including rainforests, deciduous forests, deserts, grasslands and arctic areas.

Figure 9.8 Tortoise

Reptiles are vertebrates that have dry, scaly skin. Some examples of reptiles are *snakes*, *lizards*, *crocodiles* or *turtles*. Reptiles lay **eggs**. Unlike amphibians, they can reproduce on land, far away from water. Reptile eggs are soft-shelled and often buried in a nest on the ground. Most reptiles have **four legs** (except for snakes which have none) and **claws**. Reptiles have a bony internal skeleton and are exothermic. Their **dry skin** stops water loss, allowing reptiles to be very successful in most land areas, including deserts.

Figure 9.9 Iguana

Figure 9.10 Parrot

Birds are vertebrates that have feathers. Some examples of birds are *crows*, *finches*, *ducks*, *penguins*, *parrots* and *ostriches*. Birds have a **beak** or bill, **two wings** and **two legs and claws**. The shape of a bird's wings, beak and claws often help people know the birds habitat and diet. Birds have several different kinds of feathers. **Feathers** are grown on the skin of birds and are similar to our hair. Feathers help birds stay warm, provide color, attract mates and allow most birds to fly! Birds lay hard-shelled eggs and usually raise their young in a nest. Birds have a bony internal skeleton and are endothermic. **Endothermic** animals keep a constant internal body temperature (birds stay warm because of the feathers!). Birds are found in the skies, oceans and many land areas.

Figure 9.11 Dove

Figure 9.12 Wolf

Figure 9.13 Whale

Mammals are vertebrates that have hair or fur. Some examples of mammals are *humans*, *mice*, *dolphins*, *dogs*, *kangaroos* and *bats*. **Hair**, also called fur, is made up of thousands of specialized thread-like structures found on the skin of mammals. Hair provides color and protection to most mammals. Some mammals are covered with lots of hair, like your cat. Other mammals are almost hairless, like a whale. Mammals produce milk for their young in specialized glands called **mammary glands**. After the young mammal is born, the adults usually provide a lot of parental care. Mammals have **four specialized limbs** that are used for movement. Mammals use their limbs to walk, run, hop, swim, fly and climb. Mammals usually have **large brains** and show an increased ability to learn. Young mammals learn a lot from their parents because their parents spend a long time raising them. Mammals are grouped according to how they have babies. Some mammals lay eggs, some use a pouch and still others give birth to live babies. Mammals have an internal bony skeleton and are endothermic. Mammals live in the ocean, fly in sky and live on or below the ground.

Activity

Outline animal facts. Your teacher has provided you with a blank animal outline. You may choose which group of animals you want to learn about (reptile, fish, bird, amphibian or mammal). Select your outline then use the Internet, encyclopedias or textbooks to learn facts about your animal. Write general facts about your animal group, examples of specific animals found in this group and any other interesting data. Then color your poster.

Teacher note: Animal outlines can be found in the appendix. You can just run off single sheets or blow up each image using the photocopier to create a large poster students can assemble with tape.

Science Vocabulary

Go to your school or local library, or go online to www.wikipedia.com, to access an encyclopedia. Look up the three words below.

marsupials monotremes placental mammals

What is the main difference between these three kinds of mammals?

CHAPTER 9 REVIEW

1. What is NOT a main characteristic of animals?
   A. multicellular
   B. eats other organisms
   C. has internal skeleton
   D. can move around

2. How are feathers and hair similar?
   A. they both help keep living things warm
   B. they both help living things fly
   C. they both help living things grow taller
   D. they both are made of bones

3. What are the two main groups of animals?
   A. mammals and reptiles
   B. jellyfish and fish
   C. vertebrate and invertebrate
   D. land and water

4. What is NOT a main characteristic of reptiles?
   A. internal skeleton
   B. dry scaly skin
   C. mammary glands
   D. lays eggs

5. How are amphibians different from all other vertebrates?
   A. They have feathers.
   B. They are made of many cells.
   C. They have thin skin.
   D. They have a spine.

Challenge Question

What is the main difference between the eggs of reptiles and the eggs of amphibians? Between reptiles and birds? How do you think these differences help organisms live on land?

Chapter 10
Plant Groups

Wow, we have learned a lot about grouping and the main groups of animals. But we still haven't answered one burning question: How do scientists know which living things are animals and which ones are plants? Well, I guess we have answered half the question. We know a lot about animals, but almost nothing about plants.

As it turns out, plants, like animals, are also multicellular living things. However, unlike animals, plants can make their own food. Plants use sunlight to make their food. Another unique plant characteristic is that plants cannot move by themselves. Plants can slowly grow larger or bend toward the light source. But this usually takes several days or weeks. Unlike animals, plants cannot walk, run, swim or fly over a distance in a matter of seconds.

When grouping plants, scientists use their physical structure and their life cycle, similar to the way animals are grouped. The two main groups of plants are **non-vascular** plants and **vascular** plants.

Figure 10.1 Deciduous Tree

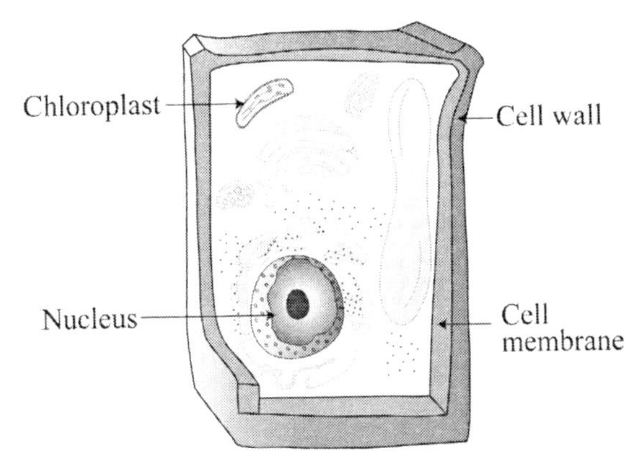

Figure 10.2 Plant Cell

You are probably more familiar with the **vascular plants**. Vascular plants are the most common type of plant found on land. This group of plants has a special group of cells, or **tissues**, that move water and food around inside the plant. Tissues are groups of similar cells that work together to complete a particular job. **Xylem** ('zī-lem) is the name for the tissues that move water inside the plant. And **phloem** ('floem) is the name for the tissues that move food inside the plant. Vascular plants usually have roots, leaves and stems. However in some plant groups, the roots, leaves and stems have special names you will learn at another time. Some examples of vascular plants include *flowering plants*, (including *deciduous trees*(trees that lose their leave in the fall), like the tree in Figure 10.1.) *conifers* (*like pine trees*), *ferns*, *cactus*, *cycads* and *ginkgos*.

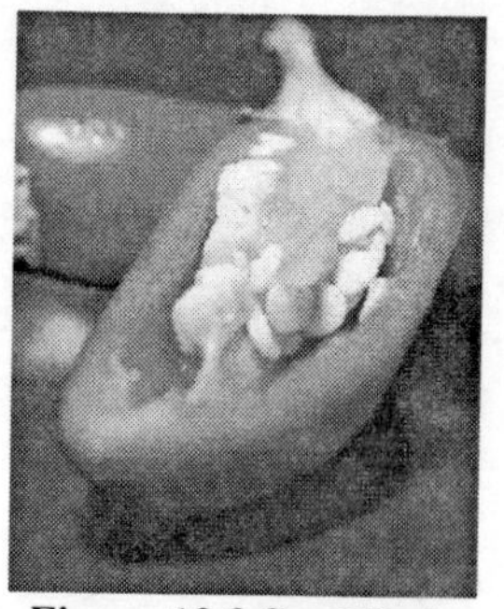
Figure 10.3 Seed Pod

Vascular plants are further divided based on whether or not they make seeds. Plants that make seeds are called **seeded vascular plants.** A **seed** is a small plant baby made from special plant parts. An entire new plant can grow from one seed. The other main group of vascular plants is called the **seedless vascular plants**. These plants make spores instead of seeds. Plant groups can be seen in the diagram below.

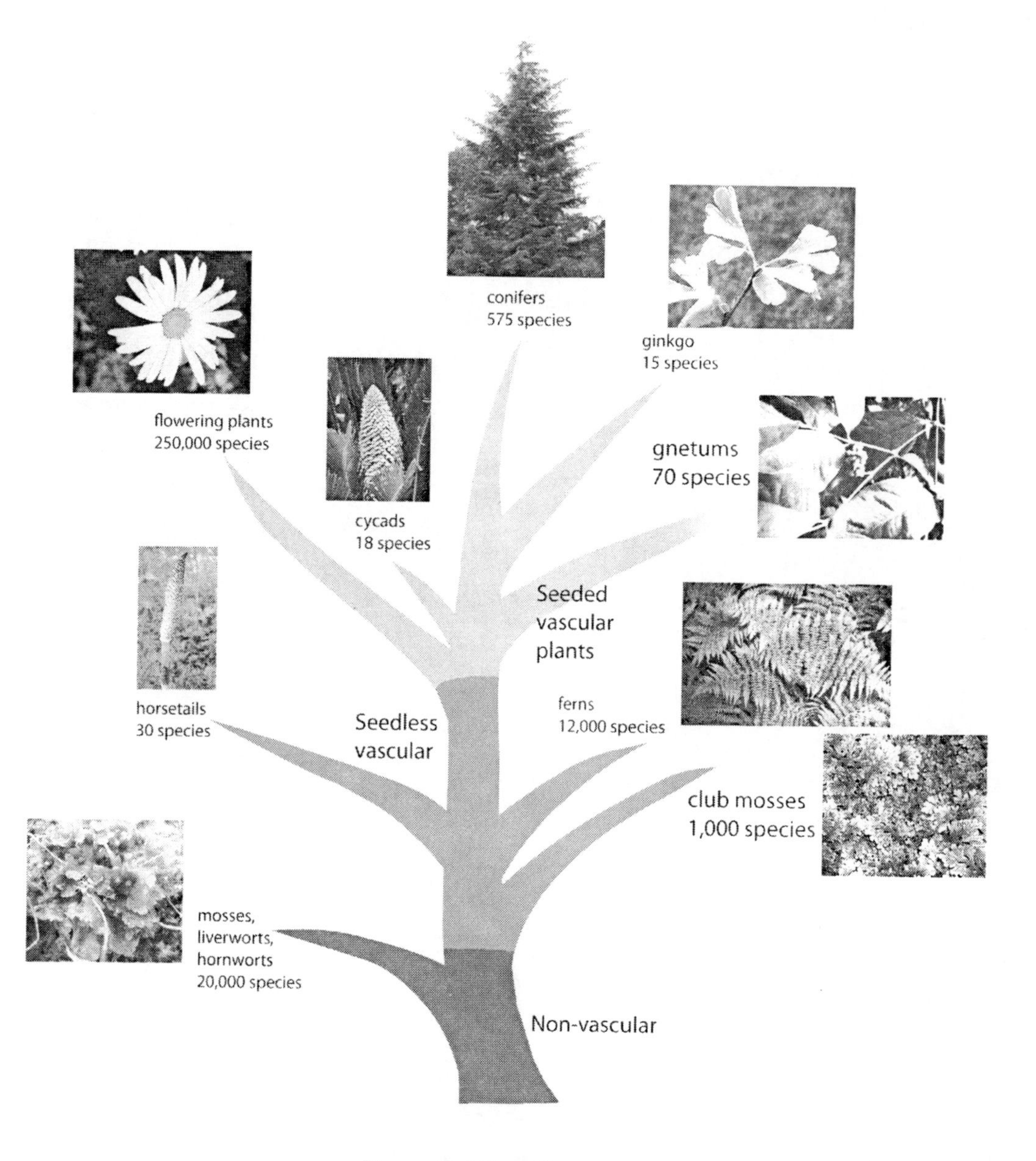

Figure 10.4 Plant Groups

In contrast to vascular plants, **non-vascular plants** are without specialized tissues. As a result, non-vascular plants usually have a very simple structure when compared to vascular plants. Because they don't have any transport tissues, non-vascular plants must live in wet or moist environments and are limited to a small size. Some examples of non-vascular plants are *mosses*, *liverworts* and *hornworts*.

Activity

Decide if each picture is a vascular or non-vascular plant. Then color and label its parts.

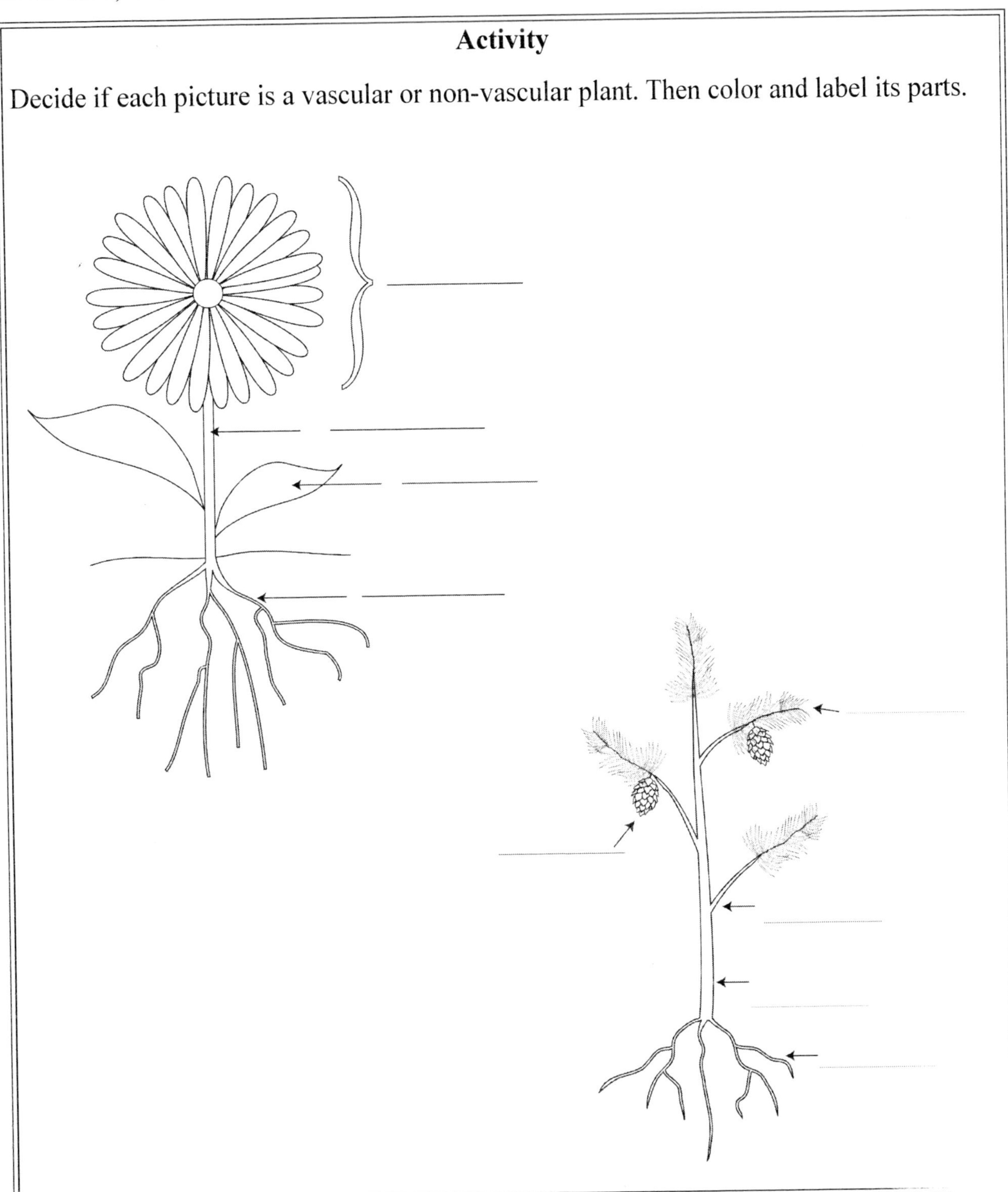

CHAPTER 10 REVIEW

1. What are the two main groups of plants?

   A. seeded and seedless
   B. vascular and non-vascular
   C. spores and seeded
   D. ferns and moss

2. How do scientists classify plants?

   A. by physical structure
   B. the color of their flowers
   C. the type of insects they attract
   D. by their chloroplasts

3. Why are mosses so small?

   A. because they make large seeds
   B. because they need sunlight
   C. because insects like to eat them
   D. because they don't have any vascular tissue

4. What does vascular tissue do?

   A. move things inside the plant
   B. keep the plant in place
   C. protect the plant from the Sun
   D. make food for the plant

5. What can plants do that animals can't?

   A. move around
   B. make their own food
   C. have babies
   D. move things inside their body

Challenge Question

What group of plants contains the most species? Why do you think this is the case?

Chapter 11
Cells

PARTS OF A CELL

Recall what we learned about microscopes in Chapter 2. Microscopes and magnifying glasses can enlarge objects (make them bigger). Making objects bigger makes it easier to learn about their structure. Like all other things, cells are made up of smaller parts. Using microscopes helps to better understand cells and their parts. The following history describes how cells were discovered with microscopes.

SCIENCE HISTORY

Figure 11.1 Hooke's Microscope

Robert Hooke was one of the first people to make and use microscopes. He lived in England from 1635 to 1703. He recorded many of his discoveries in a book called *Micrographia*. Hooke was the first person to use the word "cell." Hooke thought that looking at plant cells under his microscope reminded him of his church (specifically monk's cells).

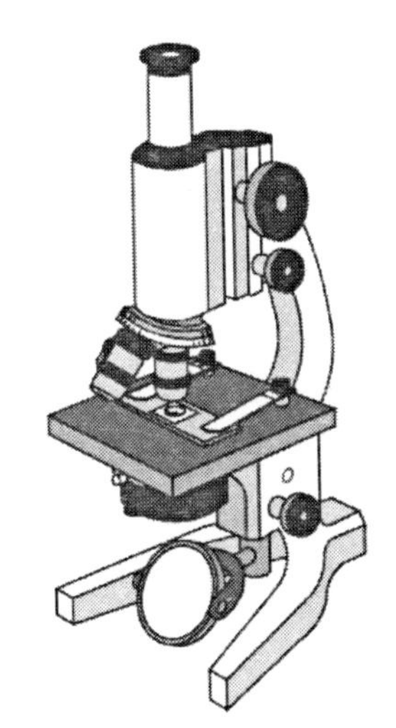

Figure 11.2 Compound Light Microscope

Anton van Leeuwenhoek was a tailor by trade and only used microscopes as a hobby. He wanted to see the tiny things that were jumping off his clients (mostly fleas and lice). Then he read *Micrographia* and was "hooked." Leeuwenhoek began working with microscopes. Leeuwenhoek made simple microscopes where one lens magnified the image. He improved Hooke's microscope design and was able to see smaller things than anyone had ever seen before. He observed things as small as bacteria under his microscopes. In fact, he was made a fellow of the important Royals Society of London for first observing single-celled organisms. Some of Leeuwenhoek's microscope designs were so good that it took another 150 years before compound microscopes could provide the same quality images. Compound microscopes use more than one lens (usually two) to magnify the image. The microscope you will most likely use in lab (also the one pictured in Figure 11.2) is a **compound light microscope**.

Activity

Look at the pictures of plant and animal cells on this page. Write down your observations about plant and animal cells in your notebook. Draw your own picture of what you see on a separate sheet of paper.

Animal Cell **Plant Cell**

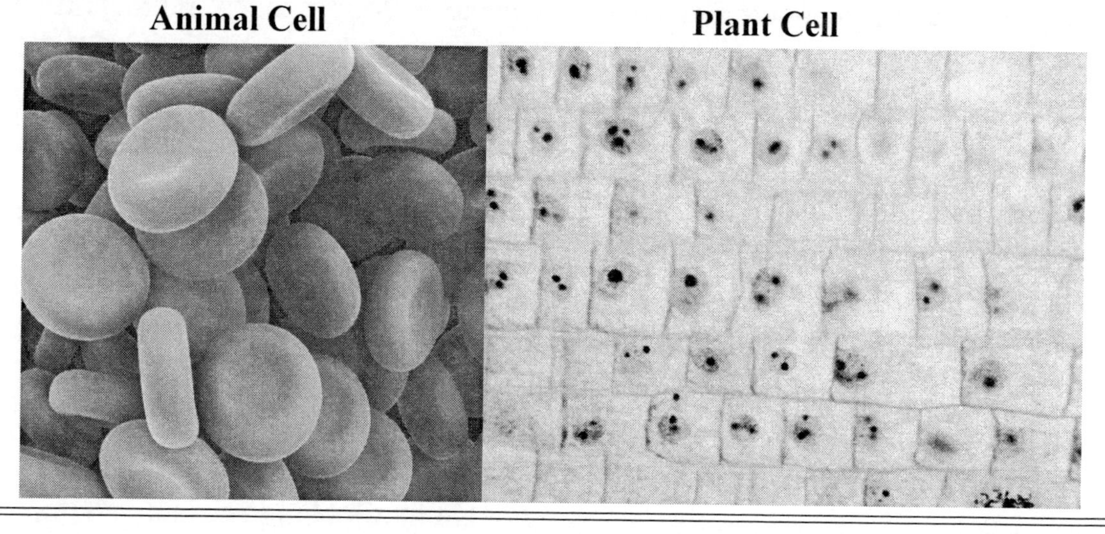

Were you able to see any differences between the different plant and animal cells? Did you notice that the plant cells looked almost like rectangles? Can you now understand why Hooke thought plant cells looked like little rooms?

PLANT CELL

Plant cells look so square because plant cells are surrounded by a cell wall. A **cell wall** is a stiff structure that separates the plant cell from the environment and gives it support. The cell wall also protects the plant cell. Located just inside the cell wall is the cell membrane. The **cell membrane** is a thin, flexible edge that also separates the plant cell from the outside environment. The cell membrane is like the cell's skin: it keeps all the cell's parts on the inside of the cell. The cell membrane also controls the flow of materials into and out of cells. You might have noticed a dark area in one part of the plant cell. This is the nucleus. The **nucleus** stores all the important cellular information, namely the DNA. The nucleus also directs all

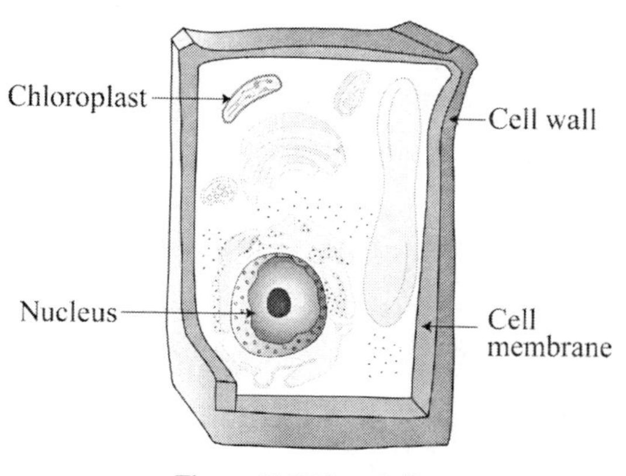

Figure 11.3 Plant Cell

activities of the cell, similar to the way your brain works. You might have noticed that not all the plant cells in the activity had a nucleus. This is because they were dead; you can only see the cell wall. Smaller structures inside the plant cell you probably didn't see were chloroplasts. **Chloroplasts** are green structures inside plant cells that capture sunlight to make food. Chloroplasts are kind of like tiny solar cells, they change sunlight into chemical energy. The food plants make is called **glucose**. But what about all that "empty" space inside the cell? Although it seems empty, this area consists of cytoplasm. The **cytoplasm** is like watery jelly and contains many important materials needed by cells.

Animal Cell

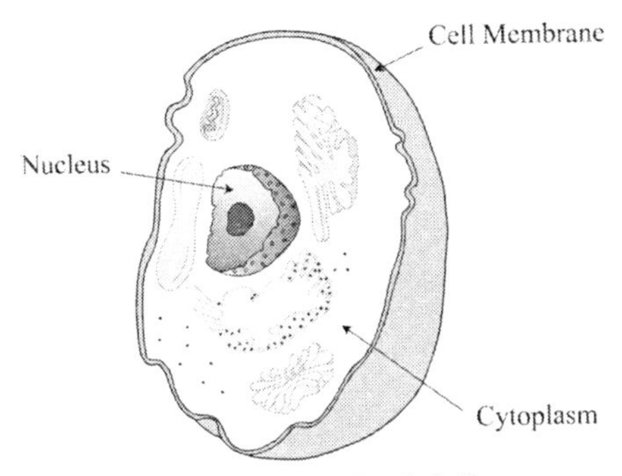

Figure 11.4 Animal Cell

In contrast, animal cells probably looked round and not square at all. That's because animal cells don't have a cell wall, instead they only have a cell membrane. In animal cells, the **cell membrane** is the only border between the cell and its environment. Similar to plant cells, animal cells also have a **nucleus** and **cytoplasm**. Animal cells do not have chloroplasts. Do you remember animals cannot make their own energy, they must eat other organisms? Mitochondria, used by both plants and animals, use the glucose made by the chloroplasts. The **mitochondria** break down the glucose into a different form the cell can easily use. Mitochondria are sometimes called the powerhouse of the cell.

Closely examine Figure 11.5. Did you notice that the space under the animal part is blank while the space under shared plant and animal parts is almost full? What does this mean? Well, in part, it means that animal cells are similar to plant cells; they share a lot of related parts. It also means that animal cells are different from plant cells because they don't have the *exact* same parts (these are things listed under plant parts).

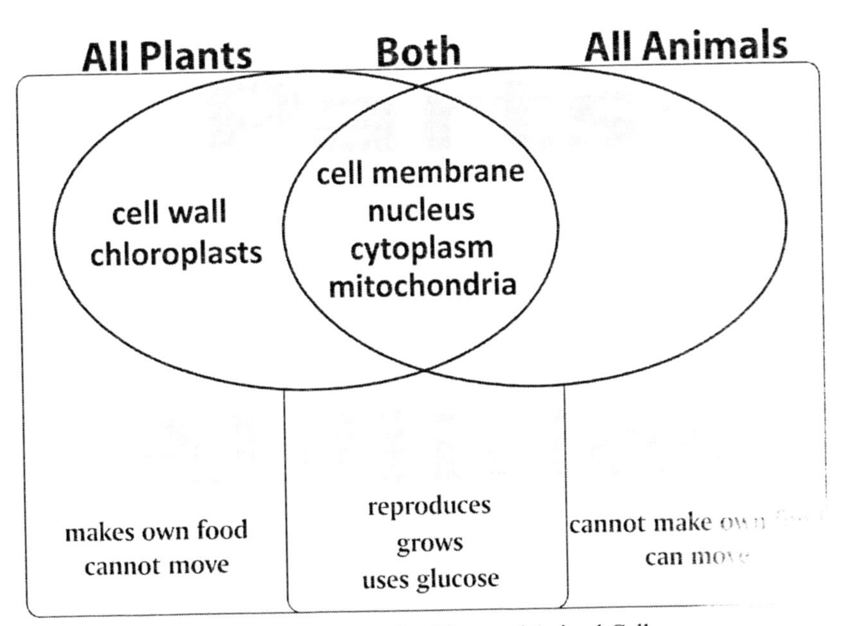

Figure 11.5 Comparing Plant and Animal Cells

Cell Differences in Multicellular Organisms

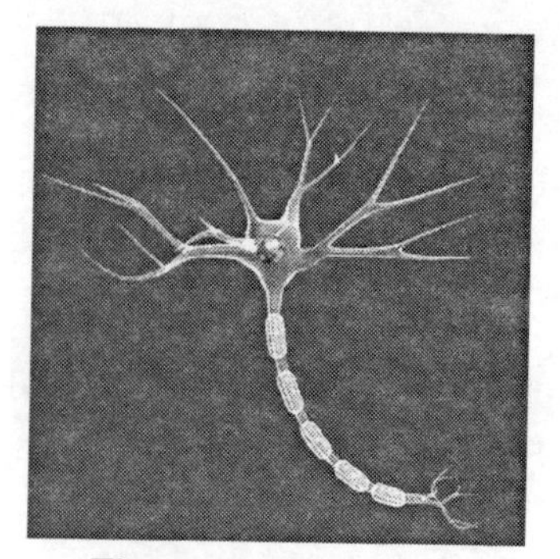

Figure 11.6 Nerve Cell

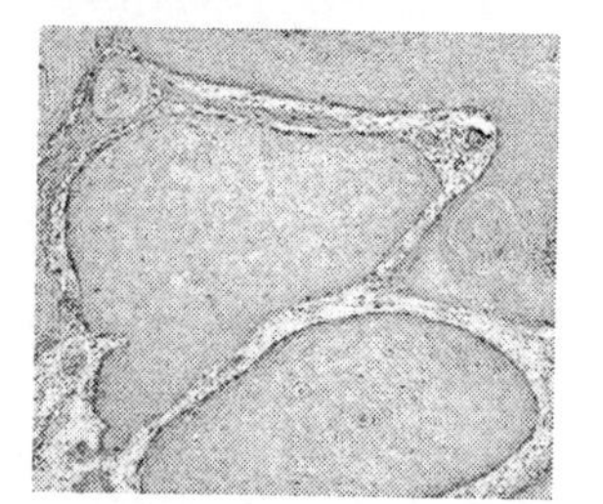

Figure 11.7 Skin Cell

Your whole body is made up of many, many different kinds of tiny cells. The cells on your skin are different from the nerve cells in your brain. The cells in your muscles are different from those in your liver.

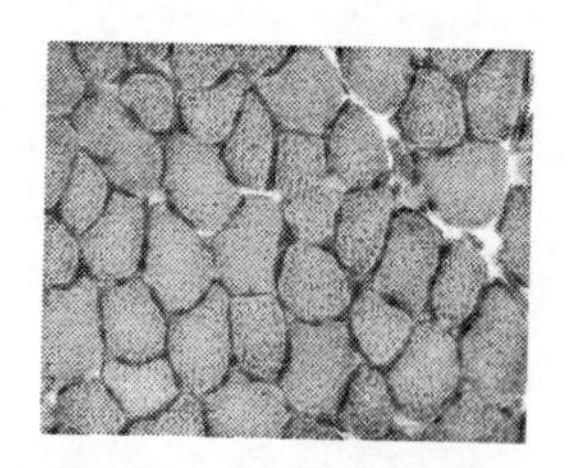

Figure 11.8 Muscle Cell

These different types of cells have changed their shape, adapting to different functions within your body. Skin cells are flat and help cover a large area. Nerve cells are long and slim to move messages over long distances.

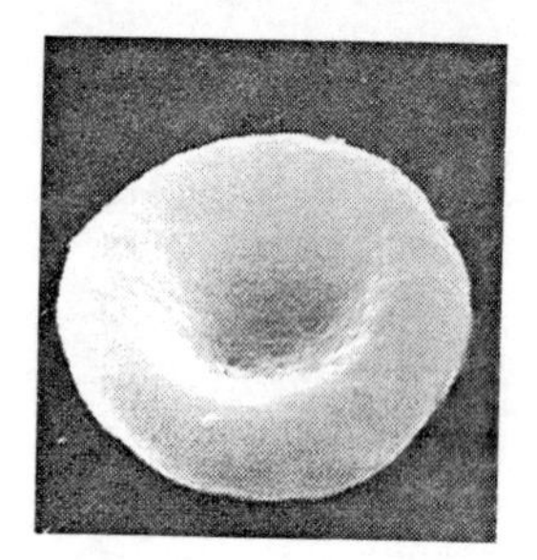

Figure 11.9 Blood Cell

Cells can group together to form tissues. A **tissue** is a group of cells that does a specific job. Sometimes cells in your body have different amounts of cell parts. Because of their high energy level, muscle cells have an increased number of mitochondria. In contrast, red blood cells have no nucleus. This allows them to more efficiently transport oxygen.

Multicellular Cells vs. Unicellular Cells

A unicellular organism is a type of living thing that only has one cell. Each unicellular organism must have all the different cell parts described earlier, all located within its single cell. It must have a nucleus (to store DNA and reproduce), mitochondria (to eat food), cytoplasm (to store things) and a cell membrane (to separate it from its environment) to stay alive.

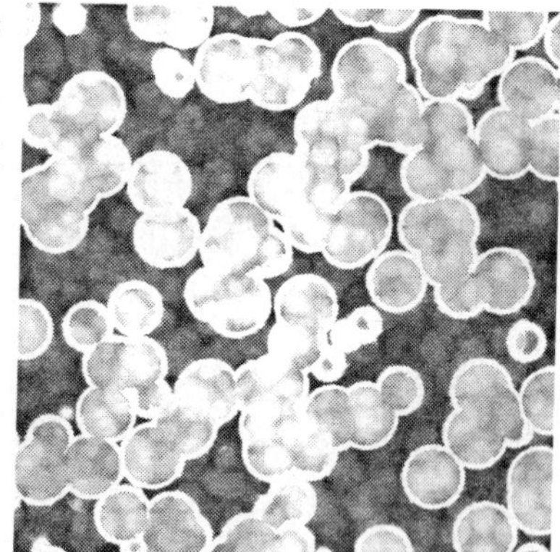

Figure 11.10 Unicellular Organism

In contrast, cells in a multicellular organism don't need to do all these things by themselves; they have other cell "helpers." Because of this they can lose some cell parts or gain others. You can think of it like this: blood is a living tissue that is part of your body, but is not alive by itself. Blood lying on the sidewalk is not a separate life form, but algae (unicellular protists) floating in a pond are separate life forms.

Activity

Label the cell parts below.

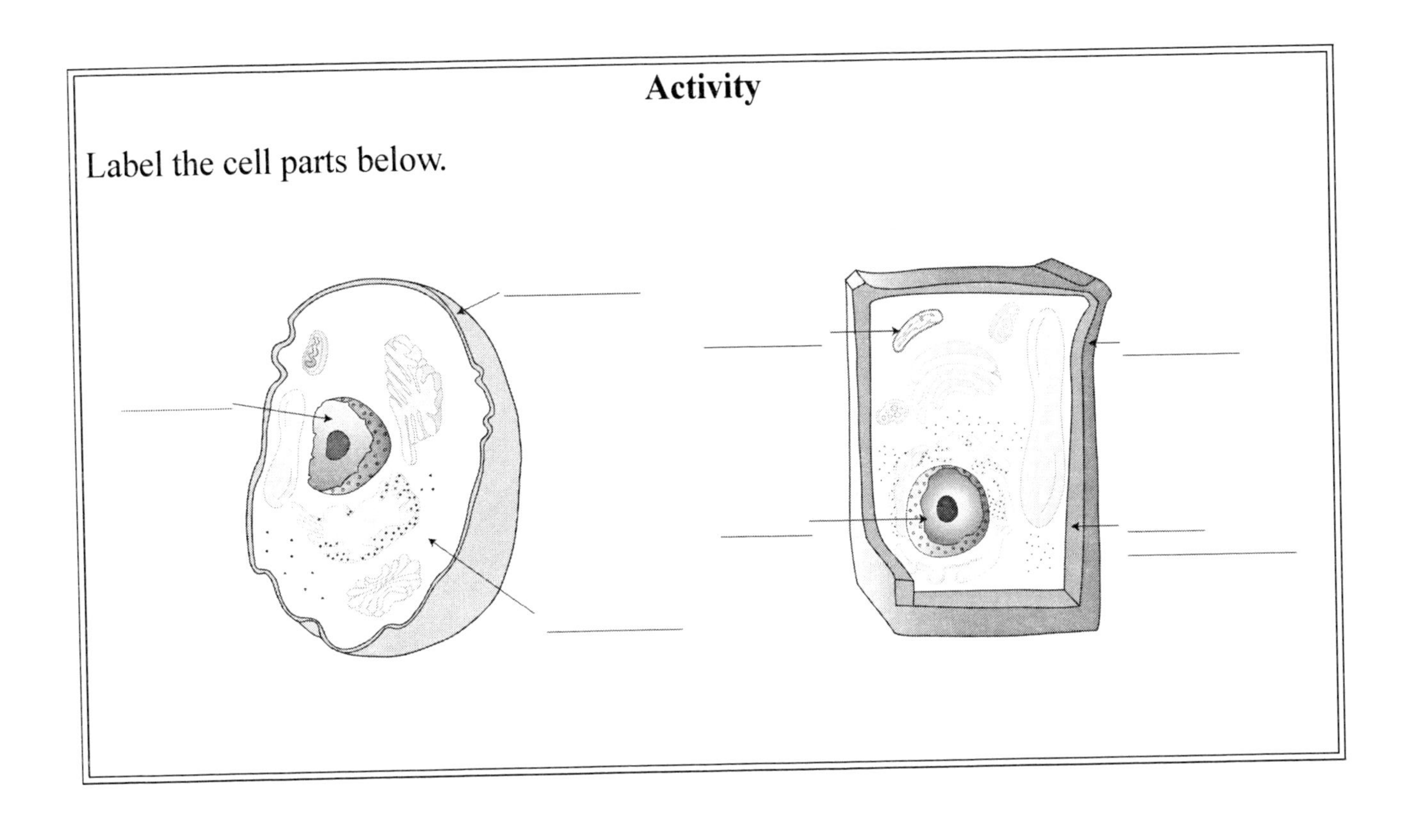

Activity

Fill in the table below.

Cell Part Name	Your Drawing of the Cell	Function of the Part
Cell Wall		
Cell Membrane		
Cytoplasm		
Nucleus		
Chloroplast		

Chapter 11 Review

1. What are groups of cells called?

   A. celloids
   B. cells
   C. tissues
   D. cell membranes

2. What do plant cells have that animal cells do not?

   A. cytoplasm
   B. nucleus
   C. cell membrane
   D. cell wall

3. What structure gives support to plant cells?

   A. chloroplasts
   B. nucleus
   C. cytoplasm
   D. cell wall

4. Which cell part helps plant and animal cells break down glucose?

   A. cytoplasm
   B. chloroplasts
   C. mitochondria
   D. cell wall

5. Which cell part stores the DNA?

   A. chloroplasts
   B. nucleus
   C. cytoplasm
   D. cell wall

Challenge Question

Look back at the activity on page 80. Why do you suppose there were no dark regions within the blood cell?

Chapter 12
I Like Your Genes

Figure 12.1 Litter of Puppies

Do you have a pet? Perhaps your neighbor has a cat or a dog. Have you ever seen a newborn litter of puppies or kittens? If you have, you might have noticed how some of the babies look very similar to the mother. Perhaps you might have noticed that you look very similar to one or both of your parents. Or maybe your grandparents have told you that you have Aunt Laurel's nose.

The familiar family likeness seen in all living things is the result of genetic traits. **Traits** are physical characteristics seen in living things. Some examples of traits include the shape of the nose, color of the eyes or height. There are perhaps thousands of different traits found in each living thing. When parents pass along traits to offspring it is called **heredity**.

Figure 12.2 Traits

Heredity is accomplished by passing along genes to children. **Inherited traits**, like the shape of your nose, are coded in your genes. **Genes** are actually the chemical "blueprints" for the physical traits. Recall from the previous chapter, the function of the nucleus of a cell. Remember that the nucleus stores the genetic information. The DNA can be thought of as the "plan" for the organism. Genes are specific smaller pieces of DNA. The genes tell the cell to make special molecules called proteins. Proteins make all the traits in your body. Basically, everything in your body from hair color to height is the result of proteins!

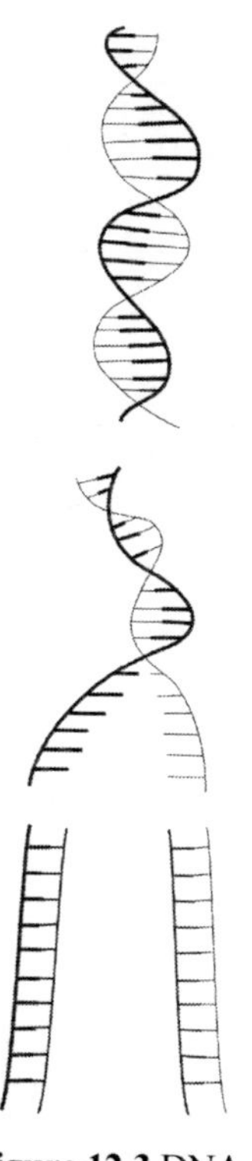

When a living thing has a baby it gives some of its DNA to the child. Some of the material inside the cells' nucleus is divided and given to the baby. A process called cell division allows organisms to donate DNA to babies. Each parent gives half of the DNA needed to the new baby. Half of your DNA comes from your mom and half of your DNA comes from your dad. This is the reason why you look like a mixture of both your parents! Your parents determine the DNA you will have when you are born.

Remember when we learned about mammals in Chapter 9? Do you remember we said that mammals learn a lot from their parents? Well this is true for you, too. You have learned a lot from your parents already. How to talk, eat and act in public are all examples of things you have learned from your parents. Have you ever wondered why people from different parts of the country have different accents? That's because they *learned* how to pronounce words from their parents. Manners, gestures, habits and ways of speaking are all things you have learned from your parents. Basically, anything you have developed the ability to do, you learned from your parents or environment.

Figure 12.3 DNA

Learned behaviors are different from inherited traits. Inherited traits were already determined at the time of your birth. Remember, things like eye color, height or shape of your nose are inherited traits. **Learned behaviors** are things you have developed the ability to do since your birth. Learned behaviors can change or become unlearned during your life. Inherited traits will never change and are coded in your DNA, in effect "hardwired" into your body.

Figure 12.4 Learned Behavior

Activity

Make a list of learned behaviors that you have that are different from your best friend's learned behaviors. Describe how you could change your learned behaviors.

CHAPTER 12 REVIEW

1. What are genes?
   A. something you wear
   B. chemical blueprints for your body
   C. behaviors and traits you learned from your parents
   D. nothing important

2. How are genetic traits passed from parents to children?
   A. by division of the nucleus
   B. through chloroplasts
   C. by learned behaviors
   D. through the food they eat

3. Your neighbor always waves hello when she sees you. What is this an example of?
   A. an inherited trait
   B. a learned behavior
   C. an innate behavior
   D. a genetic trait

4. Where does your DNA come from?
   A. half comes from your mother and half comes from your father
   B. all of your DNA comes from your mother's grandparents
   C. from your skin cells
   D. from the food you eat

5. You have blue eyes like your dad. What is this an example of?
   A. an inherited trait
   B. a learned behavior
   C. an innate behavior
   D. a fancy trait

Challenge Question

Discover and write about a learned behavior passed down from your grandparents to your parents to you. Then write about how your parents have helped shape who you are.

Your Best Friend's Dog

On a bright and ________ (adjective) spring day you go outside to ________ (verb) with your best friend Mike. When you get to Mike's house you notice something odd: ________ (adjective), muffled noises were coming from under his porch. "Mike, Mike!" you begin ________ (verb)(ing) as you start to realize what has happened. Mike's dog, Cherry, has had her puppies under Mike's ________ (noun)! Mike ________ (present tense verb) out the door asking, "What's wrong ________ (Your name)?" You explain to Mike about the noises as you both hurry under the porch. In a shallow hole, you find Mike's dog and her ________ (number) puppies. The puppies were so soft and ________ (Genetic trait) just like little cotton balls. The puppies were tiny, each about ________ (number) inches long- or about the same size as a bean bag. You sit ________ (adverb) with Mike and watch the puppies and their mother. You soon notice one puppy that is ________ (Genetic trait) and very ________ (Genetic trait) like her mother. "Hey Mike, can I take this one home if it's okay with my mom?" you ask. "Sure!" Mike replies ________ (adverb). Sure enough, your mom says okay and about eight weeks later you bring home your puppy in a ________ (noun). By now, your puppy has long ________ (Body part) and is very ________ (Genetic trait). You decide to name your puppy ________ (Name). Your mother tells you that until the puppy is ________ (learned behavior); the puppy can only come out of the ________ (Same noun as above) when someone is with him. You and Mike soon have great fun teaching your puppy tricks like ________ (learned behavior), ________ (learned behavior) and ________ (learned behavior).

Chapter 13
Surrounded by Germs

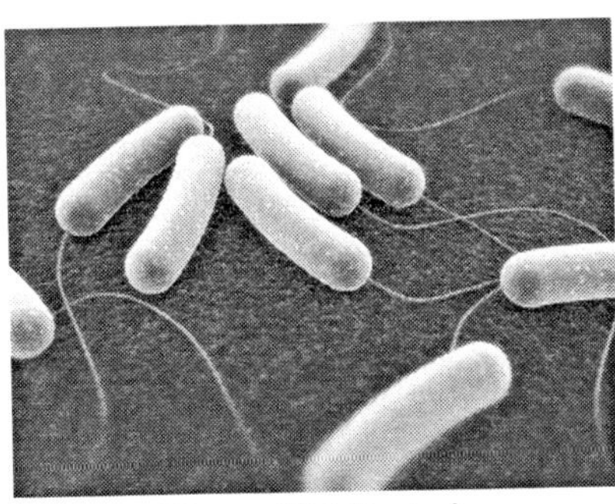
Figure 13.1 Bacteria

Have you ever heard about germs? Well in science, we call germs microorganisms. **Microorganisms** are living things that are really small. In fact, microorganisms are too small to see with your eyes, or even with a magnifying glass. They are so tiny you need a microscope just to see them!

Microorganisms are also called microbes. Microbes are usually single-celled organisms like bacteria, protists and fungi. Because they are so small, microorganisms can easily live on or inside larger organisms. In fact, microbes are found living in nearly every place on Earth.

Bacteria, fungi and protists, along with some animals, are important **helpful** recyclers in the environment. These microbes eat dead plants and animals and help keep the environment clean. A special type of bacteria called nitrogen-fixing bacteria live in the soil and on the roots of peas, beans and peanuts. This type of bacteria helps capture nitrogen from the air so plants can use it. Nitrogen is an important nutrient to both plants and animals.

Figure 13.2 Mushrooms

YUM!
BACTERIA!

Figure 13.3 A Kid Eating Yogurt

Bacteria and fungi also live on your skin. They are helpful because they eat dead skin cells and attack other harmful microbes. Some bacteria that live inside your body help you digest your food. In this way, you get more nutrients from the food you eat. Some beneficial bacteria are used to make foods like cheese, yogurt, soy sauce and sauerkraut. One type of fungi, called yeast, is used to make bread. You can also eat many different kinds of mushrooms, which are a type of fungi. (Don't eat any mushrooms you find outside, though. Some can make you really sick!)

Not all microbes are helpful; in fact some can be very **harmful**. Remember we started this chapter talking about germs. Well, germs are microbes that cause diseases in humans. Viruses, protists, bacteria and fungi can all cause diseases in humans.

Viruses are tiny things that contain their own genetic material but are not alive. In fact, viruses are smaller than the smallest known cells. Scientists need special microscopes called scanning electron microscopes or transmission electron microscopes just to see them. A scanning electron microscope magnifies 15x to 200,000x. This magnification level allows scientists to see things like never before. Using this technology, scientists have discovered many new viruses that cause human diseases. Scientists have also discovered how viruses cause diseases.

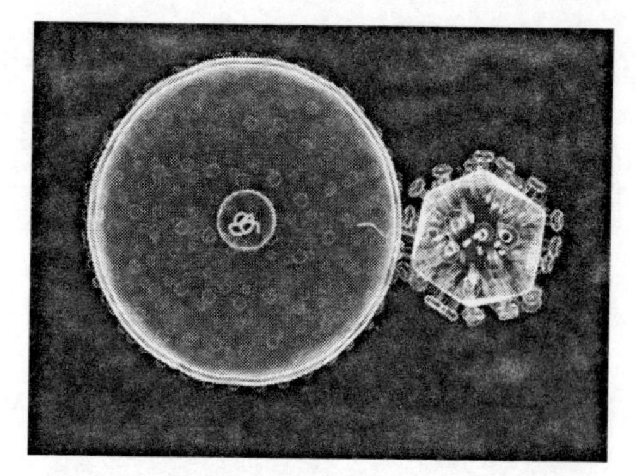

Figure13.4 Virus Inserting DNA into Cell

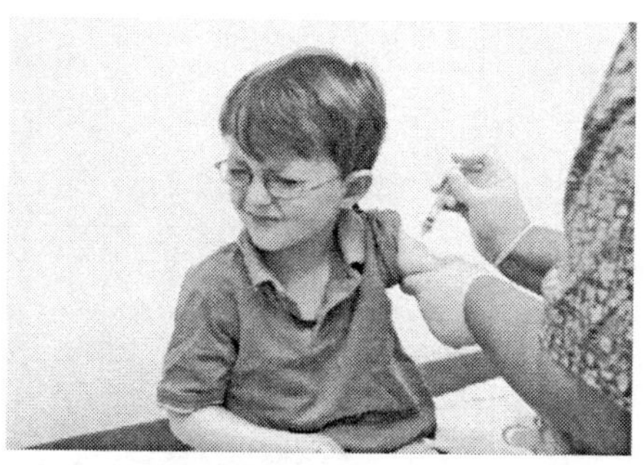

Figure 13.5 Child Receiving Vaccinations

Viruses enter the human body by putting their own DNA into human cells. Viruses can enter the human body through the skin, cuts, the mouth, eyes, nose, lungs, stomach or digestive tract. Once inside the body, the virus enters a cell. Then the virus takes over the cell and begins making more viruses. Viruses can be stopped with a **vaccine**. Do you remember getting your vaccinations before you started school? Perhaps you went to the doctor with a younger brother or sister. If you remember one of these doctor visits, you will know that a vaccine is given with a shot. A vaccine is a weak or dead virus in a special liquid. There are thousands of examples of viruses. A few diseases caused by viruses are chicken pox, the flu and measles.

Protists are a large group of microscopic organisms. These organisms can sometimes be harmful to humans or the environment.

Figure 13.6 Fish Kill from Red Tide

Protists can enter the human body through contaminated food or water or through insect bites. Perhaps the most well known disease caused by a protist is malaria. One group of protists that causes harm to the environment is the **red algae**. When there are a lot of extra nutrients in the water, the red algae reproduce quickly. This is called an algal bloom or red tide. It often happens in polluted water but not always. The rapid growth of the algae kills fish and other marine animals.

Some types of **bacteria** can hurt humans by causing fever, infection, making harmful toxins that harm tissues or causing death. The three main ways humans can become infected with harmful bacteria are cuts in the skin, inhaling the bacteria or eating the bacteria. First, harmful bacteria can enter the human body through an opening in the skin, like a cut. Secondly, harmful bacteria can enter the body through the lungs, like inhaling bacterial spores. Lastly, harmful bacteria can enter the human body through the stomach, like if you drink bad water. The human body usually fights off most infections, and the other harmful microbes, with the immune system. Bacteria can cause sinus infections, food poisoning or cavities.

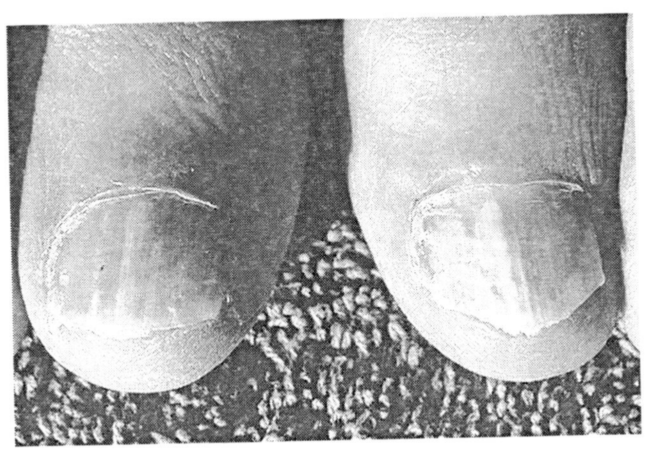

Figure 13.7 Nail Fungus

Some **fungi** live on your skin and nails. They can cause diseases by infecting the hair, skin or nails. These microbes commonly cause athlete's foot, ringworm or nail fungus.

Activity

This activity examines how quickly communicable diseases can spread. Before class began, your teacher spread "germs" (made of glow-in-the-dark powder or crushed white or fluorescent chalk) throughout your classroom. These germs were located on desks, chairs, the floor and other classroom surfaces. Under a blacklight, you can see these "germs" glow. Make some observations about who might become sick after their exposure to these germs. Collect data about possible "sick" students and create a graph showing the results of this experiment.

Teacher note: This activity is best conducted near the end of the class period. Be sure to sprinkle more chalk/powder before each new class arrives. Black out windows and draw shades to really see results. Glow-in-the-dark powder and black-lights can be purchased online or at a local hobby store.

Chapter 13 Review

1. Identify the statement below that shows how bacteria are helpful to humans.
   A. Bacteria help to digest food inside your intestines.
   B. Bacteria can cause sinus infections.
   C. Bacteria can reproduce at a rapid rate.
   D. Bacteria can create toxins on the skin that destroy tissues.

2. How do harmful protists get inside the human body?
   A. They are inhaled through the lungs.
   B. They are injected by an insect bite.
   C. They diffuse through the skin while swimming
   D. They come from infected fluid in your eye.

3. How do humans directly protect themselves from specific viral diseases like polio?
   A. with antibiotics
   B. with organic foods
   C. with vaccines
   D. with vitamins

4. Identify the term below that is another word for microorganism.
   A. vaccines
   B. antigen
   C. microbes
   D. collagen

5. Are fungi helpful or harmful to humans?
   A. They are helpful.
   B. They are harmful.
   C. They are both helpful and harmful.
   D. They do not affect humans in any way.

Challenge Question

Are viruses alive? Why or why not?

Gone Camping

While camping in the woods on a family vacation ______________ (girl's name) was thrilled with all the different vertebrate animals she saw in the woods.

The night before, she saw a mother deer with a ______________ (noun). Deer are ______________ (adjective) mammals that are herbivores and can be found in ______________ (number) parts of the country. Early this morning, she went fishing for her ______________ (noun). She caught ______________ (number) ______________ (adjective) catfish. They were around ______________ (number) inches long - or about the same size as a(n) ______________ (noun). While fishing at the pond she listened to the ______________ (adjective) chirping of the frogs. Frog eggs have a soft ______________ (adjective) shell and would dry out ______________ (adverb) if they were left out of the water.

Later on, while ______________ (verb) back to camp she found a whole field of ______________ (adjective) ______________ (adjective) flowers that seemed to shimmer brilliantly in the sunlight. Suddenly a tiny bird swooped down and was buzzing ______________ (adverb) in her ______________ (part of the body). A hummingbird! "______________ (exclamation), I've never seen one of those before," she thought. The bird zipped ______________ (adverb) to a nearby red ______________ (noun) and began to drink the nectar. Humming-birds have specialized beaks, which they use to drink the nectar from ______________ (plural noun).

Aristotle

Aristotle was a Greek ________________ (profession) who lived from ________ (number) BC until ________ (number) BC. He lived in ______________ (country) and traveled through part of Asia. He was a __________ (noun) of Plato and a teacher of _________________ (famous person).

Aristotle ______________ (past tense verb) many books in his time and his works make up a(n) _________________ (noun) of Greek knowledge. Aristotle ______________ (past tense verb) every subject ________________ (adjective) in his time. He made many intellectual contributions in physics, ____________ (school subject), logic, _______________ (school subject) and biology. Aristotle was instrumental in developing the idea that things in nature are made up of ______________ (adjective) parts. He firmly believed that by _______________ (verb) these parts we can better understand the ____________ (noun).

Aristotle was the first person to try to classify all living things found on ______________ (place). He grouped living things into two main categories: _________________ (plural noun) and ______________ (plural noun). Although his grouping system was incorrect, it did help scientists organize _____________ (noun) in a ________________ (adjective) way.

Life Science
Domain 2 Review

1. When you see a cell under a microscope, what is the first clue that tells you what type of cell it is?
   A. size of the cell
   B. color of the cell
   C. shape of the cell
   D. texture of the cell

2. Identify the statement below that explains how bacteria can be harmful to humans.
   A. Bacteria break down food in your digestive tract.
   B. Bacteria living in your mouth cause cavities.
   C. Bacteria living in on the roots of plants fix nitrogen from the air.
   D. Bacteria break down dead plants and animals.

3. Identify the cytoplasm in image below.

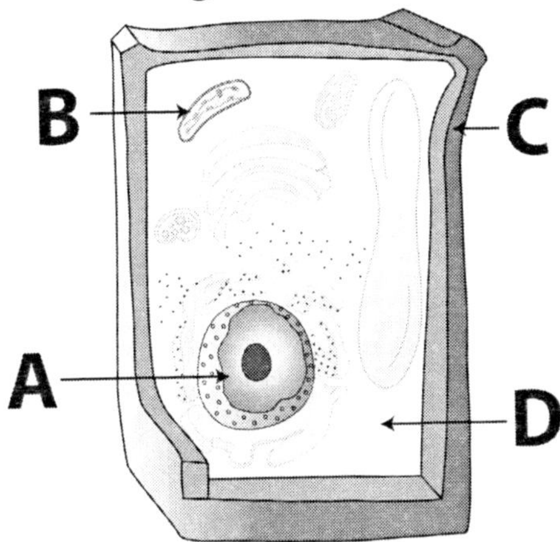

4. Identify the statement below that describes how fungi are NOT helpful to humans.
   A. Fungi are edible, providing iron and selenium to your diet.
   B. Fungi break down dead plants and animals.
   C. Fungi can make toxins that are poisonous.
   D. Fungi eat other microbes on the skin and nails.

5. Saying "excuse me" after you burp is a/an
   A. inherited trait.
   B. learned behavior.
   C. competitive behavior.
   D. example of immunity.

6. Identify the list below that sorts microbes into helpful and harmful.

A.

yeast nitrogen fixers decomposers	ringworm chicken pox measles

B.

flu yeast soy sauce	measles tooth decay food poisoning

C.

nitrogen fixers rubella yogurt	decomposers chicken pox sinus infections

D.

sinus infections ringworm athlete's foot	cavities yeast decomposers

7. Examine the image below. What type of cell is pictured here?

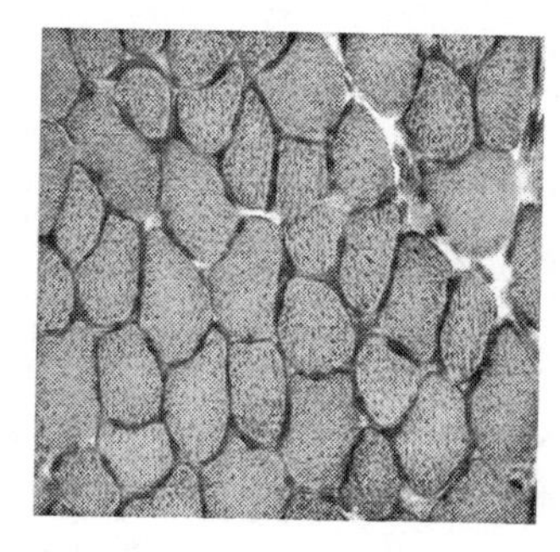

A. plant cell
B. animal cell
C. carrying cell
D. amniotic cell

8. Which group of animals has a spine?
A. vascular
B. non-vascular
C. vertebrate
D. invertebrate

9. You find a plant growing near a small creek. This plant is small and has no roots. What type of plant have you found?
A. vascular
B. non-vascular
C. vertebrate
D. invertebrate

10. The shape of your ear is determined by your
A. genes.
B. brain.
C. behavior.
D. wit.

Domain 3
Chapters 14 – 16

Chapter 14: Plate Tectonics

S5E1: Students will identify surface features of the Earth caused by constructive and destructive processes.

a. Identify surface features caused by constructive processes.

- deposition (deltas, sand dunes, etc.)
- earthquakes
- volcanoes
- faults

b. Identify and find examples of surface features caused by destructive processes.

- erosion (water—rivers and oceans, wind)
- weathering
- impact of organisms
- earthquakes
- volcano

Chapter 15: Rocks and the Rock Cycle

S5E1: Students will identify surface features of the Earth caused by constructive and destructive processes.

a. Identify surface features caused by constructive processes.

- deposition (deltas, sand dunes, etc.)
- earthquakes
- volcanoes
- faults

b. Identify and find examples of surface features caused by destructive processes.

- erosion (water—rivers and oceans, wind)
- weathering
- impact of organisms
- earthquakes
- volcano

Chapter 16: Managing Earth's Changes

S5E1: Students will identify surface features of the Earth caused by constructive and destructive processes.

c. Relate the role of technology and human intervention to the control of constructive and destructive processes.

- seismological studies,
- flood control, (dams, levees, storm drain management, etc.)
- beach reclamation (Georgia coastal islands)

Chapter 14
Plate Tectonics

THE EARTH'S LAYERS

Although the Earth appears to be made up of solid rock, it's actually made up of three distinct layers: the **crust, mantle and core**.

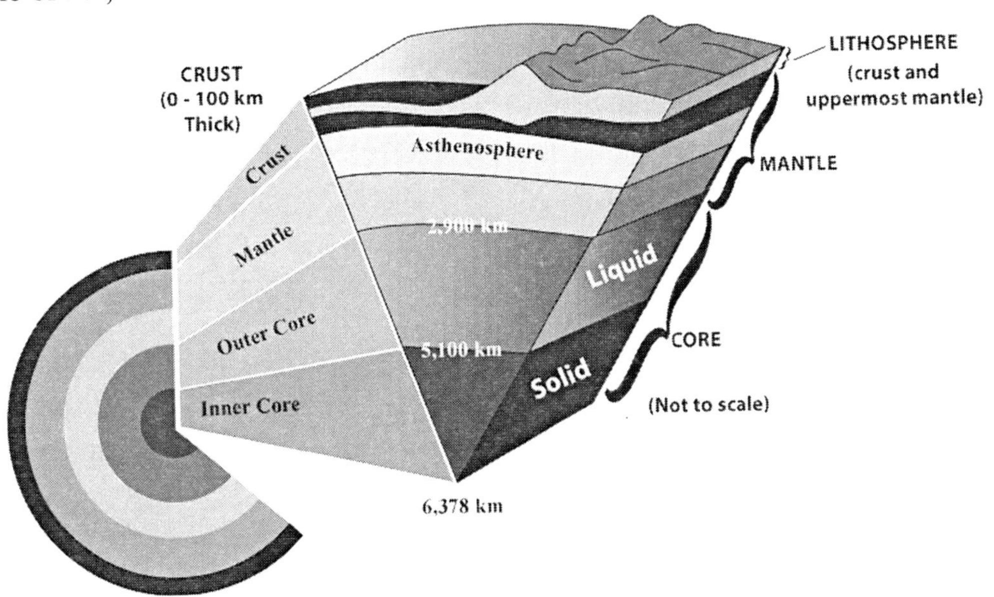

Figure 14.1 Layers of the Earth

The **crust** is the thin, solid, outermost layer of the Earth, composed mainly of basalt and granite. Basalt and granite are both igneous rocks. We'll talk more about rock types in Chapter 15. **Oceanic crust** is found beneath oceans. **Continental crust** is found on land.

The layer below the crust is the **mantle**. The mantle is much thicker that the crust, extending to about 2,900 km beneath the Earth's surface. The mantle contains more metals (like iron and magnesium) than the crust, making it more dense. The uppermost mantle interacts with the crust to form an area called the **lithosphere**. This layer of Earth is solid but brittle. Here the Earth can fracture or break.

Let's move down toward the center of the Earth to the lower mantle. The temperature rises and we reach part of the mantle that is partially **molten**. Molten means the rock is so hot it acts like plastic. It can be stretched, folded or even flow (although it's a pretty slow flow) without fracturing or breaking. (If you are trying to imagine what plastic rock would feel like, a good comparison would be Silly Putty®.) This "mid-mantle" area is called the **asthenosphere**. Going deeper into the lower mantle, temperatures rise until we reach the liquid outer core.

At the center of the Earth lies the super-dense **core**. The core of the Earth is made up of two distinct layers: a liquid outer core and a solid inner core. Unlike the Earth's outer layers with rocky compositions, the core is made up of two metals, iron and nickel. It's hard to imagine, but the core is about 5 times as dense as the rock we walk on at the surface (and a whole lot hotter!)

PLATE TECTONICS

The lithosphere is divided into 17 large plates, which are interlocked with many smaller plates. The large plates are shown in Figure 14.2. Each plate moves as an individual unit. The tectonic plates, which are made of the light, rigid rock of the lithosphere, actually 'float' on the denser, flowing athenosphere. The study of how these plates interact is called **plate tectonics**. When we talk about the "interaction" of plates, what we mean is how they move against each other. Sometimes the plates bump into each other. Sometimes plates pull apart from each other. And sometimes plates slide against one another. Each of these movements has a different effect.

Before we look at those effects, let's look a little closer at the ideas that go into plate tectonics. As with all scientific theories, it builds on the older scientific knowledge. The theories of continental drift and seafloor spreading are combined in the modern theory of plate tectonics.

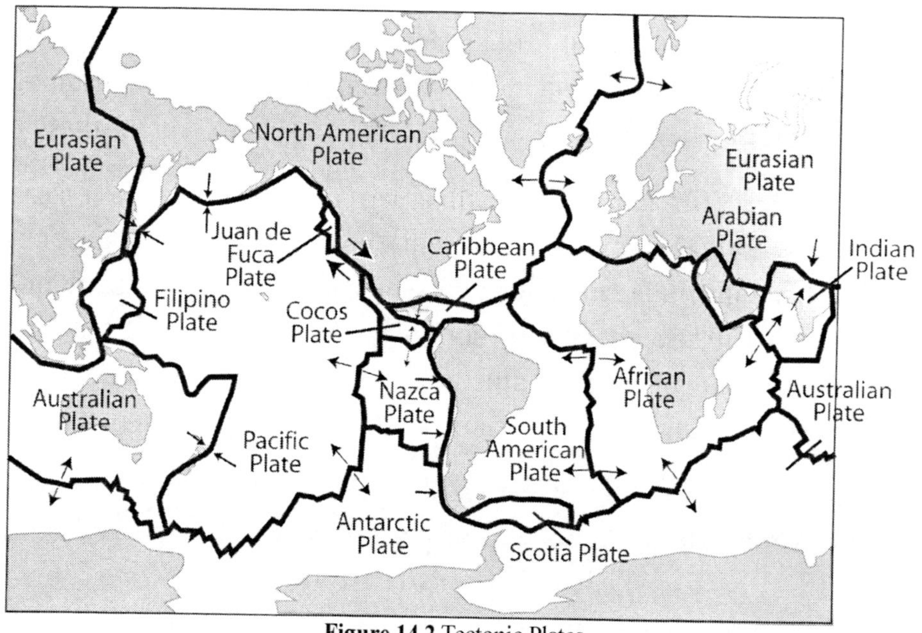

Figure 14.2 Tectonic Plates

Continental Drift and Seafloor Spreading

Scientists believe that at one time in the past (about 200 million years ago), all of Earth's continents were joined in one gigantic "supercontinent" called **Pangaea**. Imagine being able to walk from Moscow to Atlanta without having to get on a plane or boat! According to the theory of plate tectonics, the movement of the plates caused continents to drift apart in the process known as **continental drift**.

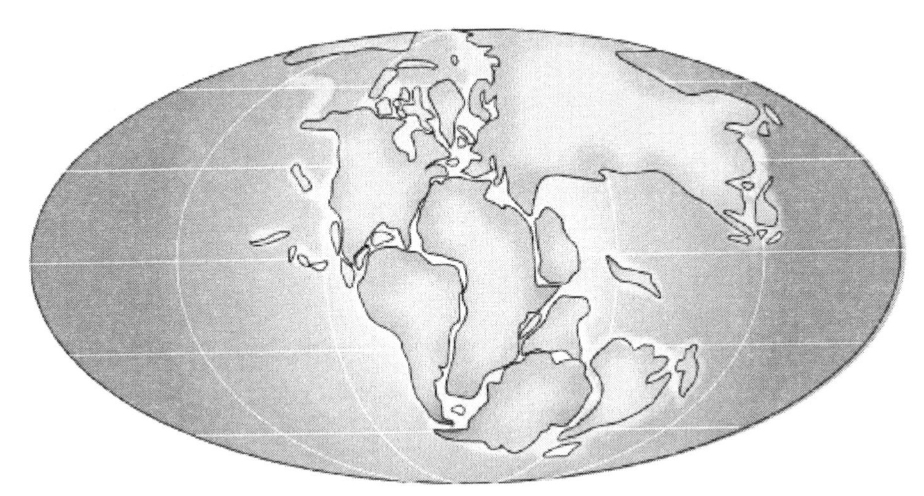

Figure 14.3 Pangaea

But how do the plates move? Hot thick, liquid rock passes through the mantle, in currents like the ocean (but much slower). These massive currents rotate in circular patterns, carrying heated rock upward. The molten rock soon reaches the crust. When the currents bring magma up under the thicker continental crust, it doesn't usually break through. Instead, it changes direction and moves underneath the crust. This drags the Earth's plates along with the underlying material. As the heated material cools, it sinks back into the mantle. New molten rock rises to continue the process.

Where the crust is thin, as oceanic crust is, the molten rock may break through a crack. The cracks are places where plates are moving away from each other. The hot **magma** (liquid rock) seeps up through the crack and meets ocean water. The water quickly cools the magma and it hardens to a solid. New ocean crust has been formed. This process is called **seafloor spreading**. The underwater landform that results is called a **mid-ocean ridge**. Mid-ocean ridges contain young rock, whereas other areas of the ocean floor contain older rock. The oldest rock is found on the continental crust.

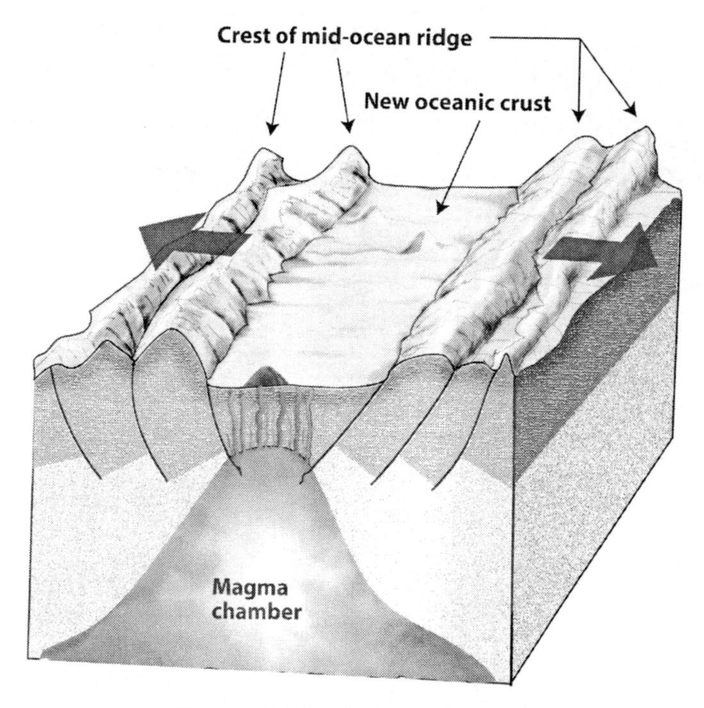

Figure 14.4 Seafloor Spreading

Constructive and Destructive Processes

The movement of the tectonic plates affects the features found on the crust. We call features of the Earth **landforms**. Geologists describe changes in landforms as constructive or destructive. For instance, the process of seafloor spreading adds more crust on the ocean floor. It also creates mid-ocean ridges. Therefore, seafloor spreading would be a constructive process. A **constructive process** contributes to the formation of new landforms, or addition to existing ones.

Other processes are destructive. A **destructive process** contributes to the destruction of landforms. A good example of a destructive process is when tectonic plates bump into each other. The result is that one plate slides under the other. Anything on the surface is sucked into a deep trench. When this happens on the ocean floor, it is called a **deep sea trench**. For instance, the Marianas trench in the Pacific Ocean is the deepest known place in any ocean, at nearly 11,000 meters below sea level.

Keep in mind that destructive and constructive processes occur on the continental crust, too. We will discuss these processes, and the landforms that they affect, in the next chapter.

Activity

The following figure contains pictures of oceanic landforms. Use it and their descriptions to decide whether a constructive process or a destructive process created the landform. Don't worry if you are not sure. Make a guess and discuss your reasoning. You will learn more about these processes in the next chapter. (And, also, here is a hint: sometimes landforms are the consequence of both constructive AND destructive processes)

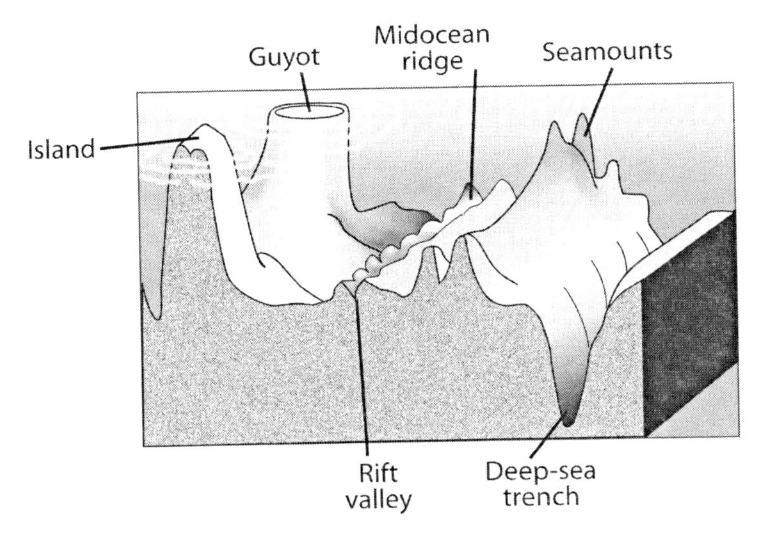

Seamounts: Underwater mountains that do not reach the surface. Seamounts are usually formed when a tectonic plate moves over a volcanic hotspot deep within the Earth's mantle. Seamounts may have once been volcanoes.

Guyot: A flat topped seamount. These were seamounts that once rose above the surface of the water. Their tops were worn down or knocked off by wave action. Like seamounts, guyots may once have been volcanoes.

Island: There are several kinds of islands. One kind is formed by the same process that forms seamounts, but the volcanic activity is stronger. The greater force forms a seamount that rises out of the water.

Rift Valley: The natural result of ongoing seafloor spreading.

CHAPTER 14 REVIEW

1. Which layer of the Earth is the most dense?
   A. the crust
   B. the lower mantle
   C. the upper mantle
   D. the core

2. What two kinds of crust are there?
   A. lithospheric and asthenospheric
   B. silicon-based and iron-based
   C. continental and oceanic
   D. outer and inner

3. The ancient supercontinent, Pangea, once contained
   A. all of the Earth's tectonic plates.
   B. all of the Earth's crust.
   C. all of the Earth's continental crust.
   D. all of the Earth's oceanic crust.

4. A constructive process can occur
   A. when plates bump into each other.
   B. when plates move away from each other.
   C. when plates break apart.
   D. only on the sea floor.

5. Which of the following are produced by a destructive process?
   A. seamounts
   B. islands
   C. midocean ridges
   D. deep sea trenches

Challenge Activity

Tectonic plates can interact in one of three ways. Think of a way to draw each interaction. You want to illustrate the movement of the plates with your drawing.

GEORGIA 5 CRCT SCIENCE

Chapter 15
Rocks and the Rock Cycle

ROCKS AND THE ROCK CYCLE

There are three types of rocks: igneous, sedimentary and metamorphic. **Igneous** rocks are formed from magma beneath the Earth's surface. **Sedimentary** rocks are formed from deposited and compressed **sediments** and are found in layers. **Metamorphic** rocks are rocks that have been changed by heat and pressure.

WEATHERING

Not even rocks last forever on the Earth's surface! When a rock is broken down into smaller pieces it is called **weathering**. A rock **weathers** in response to changes in its environment. A rock buried in the Earth can remain the same for millions of years, but when exposed to wind and water at the Earth's surface, it will weather. Weathering can occur by three different methods: **mechanical weathering**, **chemical weathering** and **biological weathering**.

Mechanical weathering is a process where rocks are physically broken into smaller pieces by wind, water, ice or heat. The common product of mechanical weathering is **silt**, a form of very finely-ground rock.

Examples of this type of weathering are everywhere. For example, the Grand Canyon is a channel cut through rock by the powerfully surging waters of the Colorado River over a period of many millions of years.

Figure 15.1 Channel
Grand Can

Freezing and thawing cycles also weather rock mechanically. If you ever travel up the Eastern Seaboard to a northern state like New Jersey or New York, you find that the roads become bumpier and road maintenance more common. This is because the winters are colder up north. Why should that matter? Water gets into the cracks and crevices of the concrete and asphalt that roads are made of (concrete and asphalt are just processed rock). When the water freezes, it expands. The ice crystals push against the solid rock, weakening its structure. When it warms up again, the ice crystals melt and the
released — but the rock structure is weakened even more. Each time the rock
thaws, it cracks a little more. Finally, the road is rubble.

Chemical weathering is a process where minerals within the rocks are broken down by removing or altering elements that make up the minerals. The most common form of chemical weathering is the result of carbon dioxide (CO_2) in the air. When the carbon dioxide combines with rain, they undergo a chemical change. They form a weak acid. This acid dissolves limestone formations. Different acids are produced by volcanic eruption and pollution. These acids affect other kinds of rock. Small rocks have more surface area exposed than larger rocks, so they weather more quickly. The end result of chemically weathered rock is **clay**.

Biological weathering is weathering caused by living organisms. It can occur by either mechanical or chemical means. For example, burrowing animals or plant roots can break up soil and rocks mechanically. Lichens chemically weather the rocks they live on by secreting enzymes that remove nutrients from the rock and reduce it to soil.

Figure 15.2 A Weed Biologically Weathers the Concrete

EROSION

Erosion is the transport of soil or rock by water, ice or wind. Erosion is the *movement* of weathered rock, and NOT the actual weathering itself. So, streams and rivers not only weather rocks by carrying water past the rock. They also transport the fragments downstream (erosion) by carrying them in the water or rolling them along the bottom. This is a form of water erosion. For instance, a river can alter its course, cutting a new channel. In the process, soil is picked up, then deposited as sediment down river.

Water erosion also occurs along ocean shorelines. At the shore of the ocean, waves weather and transport rocks along the beaches. Over time, beaches erosion either shrinks the beach or moves it.

Disasters also cause erosion. Think of hurricanes, floods and tornadoes. The December 26, 2004 **tsunami** (huge sea wave) in Southeast Asia caused tremendous loss of life and property. It also eroded beaches and land masses, moving sand and soil along the coasts of many countries.

Wind erosion occurs all over the Earth but is most noticeable in desert climates, where water is scarce. The wind picks up small pieces of rock fragments and carries them along until they reach an obstacle, such as a hill. The fragments, pushed by the wind, act as a sandblaster, eroding the hill slowly over time. The wind also causes erosion by scooping up large areas of loose soil and transporting it to another location.

One example of wind erosion in our own nation's history is the **Dust Bowl**. As settlers moved into the central United States, they removed the vast plains of grass to clear the area for farming. Poor farming techniques, combined with a drought, left the soil dry and loose. Wind erosion moved the dusty soil and destroyed much of the Great Plains.

Figure 15.3 The Dust Bowl

CONSTRUCTIVE AND DESTRUCTIVE PROCESSES

Weathering and erosion are called **destructive processes**. They each are responsible for tearing landforms down or changing their structure. On the other hand, weathering creates sediment. Sediment is eroded and deposited somewhere. This **deposition** of sediment is a **constructive process**. It contributes to the formation of landforms, or their re-shaping. Deltas are formed by deposition of sediment. Figure 15.4 shows the Ganges River Delta. This is the largest river delta in the world. You can see from Figure 15.5 that the Ganges is a very long, wide river until it reaches its end. Nearing the bay of Bengal, it splits off into many tributaries (small rivers). Each of these erodes soil. The eroded soil is carried by all the tributaries to the bay where it is dumped. Sand dunes at the beach or in the desert are also examples of erosion.

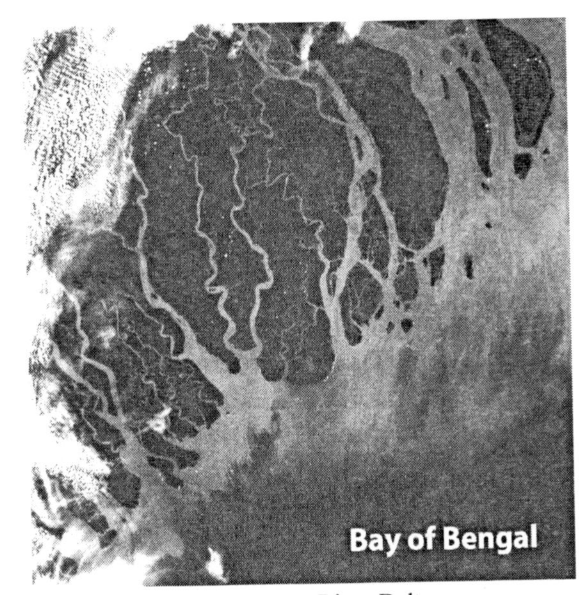

Figure 15.4 Ganges River Delta

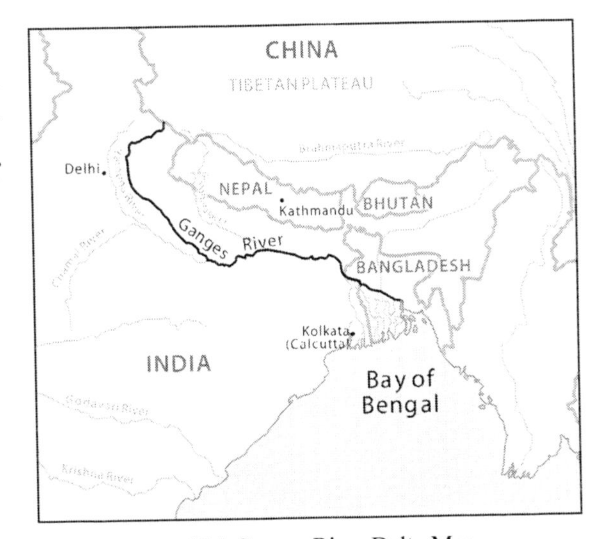

Figure 15.5 Ganges River Delta Map

Volcanic Activity

A **volcano** is a mountain fromed from **lava** and rocks made from materials that have emerged from inside the earth. **Magma** is a combination of liquid rock material and dissolved gases deep within the Earth.

Figure 15.7 Volcanic Vent

Magma is generated when solid rock from the crust is forced down into the Earth's asthenosphere at a **plate boundary**. Volcanoes typically occur where plates collide.

Magma reaches the Earth's surface through an opening called a **vent** that begins deep inside the Earth. When magma reaches the Earth's surface, it is called **lava**. The lava flows out of the vent and as it hardens, it builds up and forms a mountainous structure. When the mountain is formed, the vent still has a way to release magma — through a crater. The **crater** is the hole in the top of the volcano. A cross-section of a volcano is pictured in Figure 15.7.

So, volcanoes are actually landforms made from a constructive process. They are mountains built up over time by the hardening of lava after the volcano erupts.

Activity

Describe each landform below. Look carefully and think like a scientist. What features are shown in the picture? What process or processes formed it? Which of the processes are constructive and which are destructive?

THE ROCK CYCLE

Scientists use the **rock cycle** as a model to describe rock changes on the Earth. These changes show how each type of rock is formed. Some of the processes such as weathering, erosion and deposition occur at or near the Earth's surface. Other processes, such as melting and increased heat and pressure occur deep below the surface. The rock cycle can begin with igneous rocks, created when magma cools and solidifies. Igneous rocks on the Earth's surface can undergo weathering and erosion, which creates sediment. Most sediment deposits are in oceans, but some ends up in river flood plains, desert basins, swamps and dunes. Sedimentary rocks develop from sediment (weathered and eroded rock) that becomes compacted and cemented together. Sedimentary and igneous rocks within the Earth can also be subject to heat and pressure. That can then create metamorphic rock!

It doesn't always work this way, though; since it is a cycle, we could choose to begin anywhere! Let's start again with sedimentary and metamorphic rocks. Sometimes sedimentary and metamorphic rocks are exposed to weathering and erosion and create sediment. The cycle also shows that it is possible for metamorphic, sedimentary and igneous rocks within the Earth to melt, forming magma. When magma is sent to the Earth's surface through volcanic action, the cycle begins again.

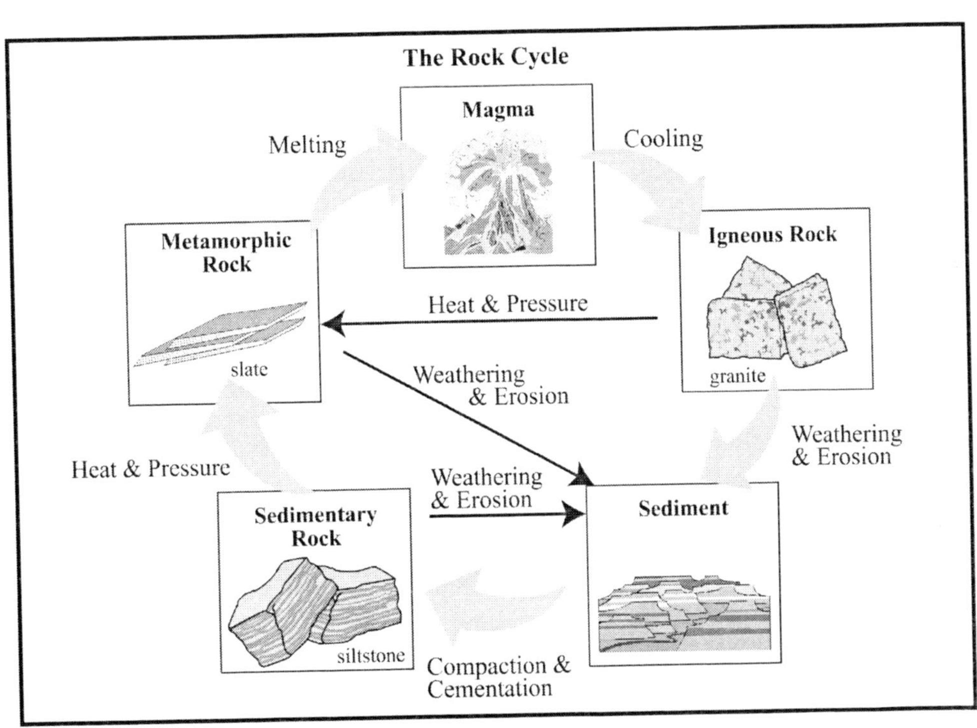

Figure 15.6 The Rock Cycle

CHAPTER 15 REVIEW

1. Wind causes weathering and erosion because it
   A. transports chemicals.
   B. transports particles.
   C. transports microbes.
   D. causes rapid cooling.

2. Which of the following landforms is the result of a process that is both constructive and destructive?
   A. sand dune
   B. canyon
   C. the Dust Bowl
   D. waterfall

3. Water weathered the Grand Canyon
   A. mechanically.
   B. chemically.
   C. biologically.
   D. automatically.

4. The hole in the top of a volcano is called a
   A. vent.
   B. crater.
   C. magma boundary.
   D. lava valve.

5. Metamorphic rock is rock whose structure has changed. What factors change the rock?
   A. heat and pressure
   B. just heat
   C. weathering
   D. erosion

Challenge Activity

Given all you know about constructive and destructive processes, describe an earthquake. What landforms could be created by an earthquake, both on the surface and under water?

Pokemon Can Be Destructive

A large valley was set between high mountains on either side. Part of the valley formed a small lake. And inside that lake, there lay a Pokeball. Two trainers fought over who would get the Pokeball. And you know what that means… the Pokemon were battling again.

The trainer ____________ (Boy's name) was a nice guy. The trainer ____________ (Boy's name) was a nasty guy. Their Pokemon quickly became a whirling blur of fur and ____________________ (body part, plural). You could hardly tell which was which! One Pokemon spit __________ (liquid) at the other. It missed and hit the mountainside. The mountain ____________________ (past tense destructive process) to form a ____________ (landform). In the split second that it had, the other Pokemon breathed fire. All the ______________ (type of plant, plural) in the valley burned. Both Pokemon launched into the air, still fighting.

The valley was changing. ________________ (Landform, plural) appeared that were not there before. The air became ___________ (adjective). Both trainers knew that other trainers and Pokemon had arrived to join the fight.

On the __________ (landform), a __________ (adjective), _____ (noun)-type Pokemon was seen hurling ____________ (noun, plural) at a ___________ (adjective) _____ (noun)-type Pokemon. Two __________ (adjective) _____ (noun)-type Pokemon ___________ (past tense action verb) off of a __________ (landform). A big pile of struggling Pokemon of all types formed next to the __________ (landform). ____________ (Plural noun) and electricity and __________ (plural noun) flew from the mound of Pokemon.

A _______ (noun)-type Pokemon seemed to fly past the trainers' eyes every moment. This many Pokemon in such a small area were making a huge mess. The valley was starting to ________________ (chemical change). The mountaintops ____________ (physical change). The ________ (landform) began to

__________________. The rocks along the river changed from __________ to
(destructive process) (type of rock)

__________. Even the trainers' ______________ began to ______________.
(type of rock) (type of shoe, plural) (physical change)

Then, suddenly, a ________-type Pokemon reached the center of the lake. His trainer
(noun)

____________ and placed the Poke-ball on his __________. He shouted "I am the
(past tense verb) (body part)

__________est trainer in the whole world!" The trainer opened the Poke-Ball. Out popped
(adjective)

a ________, ______-type Pokemon. It was eating __________. It healed all the other
(adjective) (noun) (food)

Pokemon, and they no longer wished to fight. It gave everyone in the valley __________
(food)

and a __________. The valley changed back to normal. It appeared that disaster had
(toy)

been avoided...this time.

Challenge Questions:

1. This MadLib asked for a lot of different landforms. Which of the landforms that you used can be found in Georgia?
2. One part of the MadLib asks for a ____________________. Jen used
(past tense destructive process)
"eroded". Steve used "deposited". Which one used the term that was asked for?

Chapter 16
Managing Earth's Changes

Preparing for the Possibilities

Changes are continually taking place in the crust of the Earth and in the atmosphere just above it. When the changes cause events like hurricanes, erosion and earthquakes, loss of human life and property are the result. Humans try to develop technologies to prevent these events. We hope to protect people and property from the worst effects of natural disasters. These efforts function in one of three ways.

(1) *Prediction* of the event. This gives us an early warning, and allows people to move to safety.

(2) *Preparation* for the event. Property can be secured and first aid or food supplies distributed.

(3) *Reconstruction* after the event. We cannot stop nature. No matter how much we predict or prepare, we will almost always have to rebuild after a natural disaster.

Let's look at three different areas of Earth management.

Flood Control

We like to be near water in order to swim and play. Also, all living things (like us) require water to survive. So, living near a river, lake or ocean seems like a good thing, right? Well, yes and no. Water can also become deadly as in the case of flooding.

Figure 16.1 Patio Drain

Usually when it rains, water is absorbed into the soil or evaporates into the air. That's how the water cycle works. But when a lot of rain falls in a short period of time, **floods** can occur.

Engineers have developed some ways to control the flow of water in cities during storms. One of the most important of these is the **storm drainage system**.

Figure 16.2 Storm Drain

Storm drains are separate from the **sewage system** (that's where household waste goes). A storm drainage system begins above ground. Channels and gutters are cut into roads and landforms to make paths for the water. These paths lead water to the storm drains. You've seen these, even if you don't really notice them. They are covered by grates and located at the side of the road (Figure 16.2). Some small inlets may also be found in the middle of patios, parking lots or in yards (Figure 16.1). These drains lead to large underground pipes that eventually empty into **reservoirs** (man-made lakes built to hold water reserves), larger bodies of water or the ocean.

When it rains very hard or for many days, the soil becomes saturated (filled) with water. The possibility of flooding is increased when the soil is waterlogged, because the ground cannot absorb any more water. During these times, it is especially important to have a good storm drainage system, to move the water away from homes and off of streets.

When a city is built on a floodplain, there is a high risk of flooding. **Floodplains** are flat, low-lying areas next to rivers. Because the community does not sit high up on a hill or ridge, floodwaters reach homes and businesses easily.

Engineers have developed structures to control the overflow of rivers and lakes in floodplains. The easiest way is to build a levee. A **levee** is a barrier built on the banks of a body of water. Man-made levees are usually made of dirt, sandbags or concrete. Natural levees often form as the consequence of erosion and sediment deposits. Levees keep water from overflowing the banks of the river or lake.

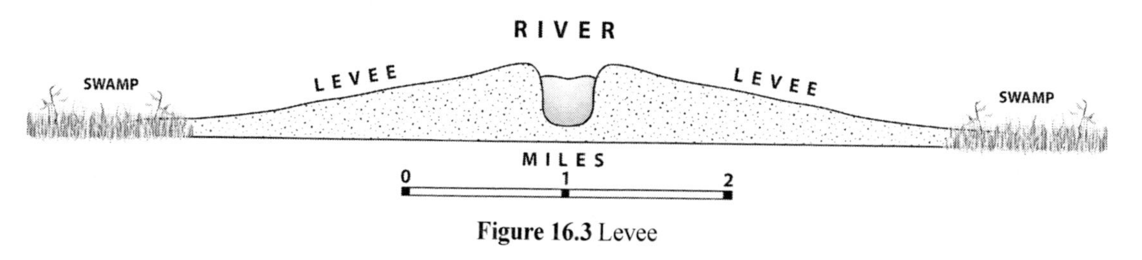

Figure 16.3 Levee

Engineers can calculate how high and how strong a levee needs to be. To do this, they use a computer model. The **computer model** is a program designed to predict the consequences of a set of actions. They put information into the model, like how low the floodplain is and how likely storms are in the area. The model says how high the levee should be, based on the input information. These calculations are important: if the levee fails, water flows into the community.

You may remember news reports of the awful flooding in New Orleans after Hurricane Katrina. This was mostly due to the failure of the levees. Heavy rains filled Lake Pontchartrain and swelled the Mississippi River. Finally, the water got too high, and the levees were overwhelmed. The water flowed over the edge of the levees and 80% of the city was submerged.

Another structure that can help with flooding is a **dam**. Dams are different than levees. While levees are used to keep water from flowing into certain areas, dams either control the flow of water downstream or stop it. Dams create large reservoirs at the place where the water is stopped.

A dam can also be useful in another way. The water that moves through it can be used to turn the turbines of an electric generator. These dams are actually hydroelectric power plants. **Hydroelectric** means that the electricity is generated by the movement of the water. A picture of an actual dam in Georgia is shown in Figure 16.4. Fig. 16.5 shows the inside of the hydroelectric power plant.

Figure 16.4 Carter's Reservoir and Dam in Chatsworth, GA.

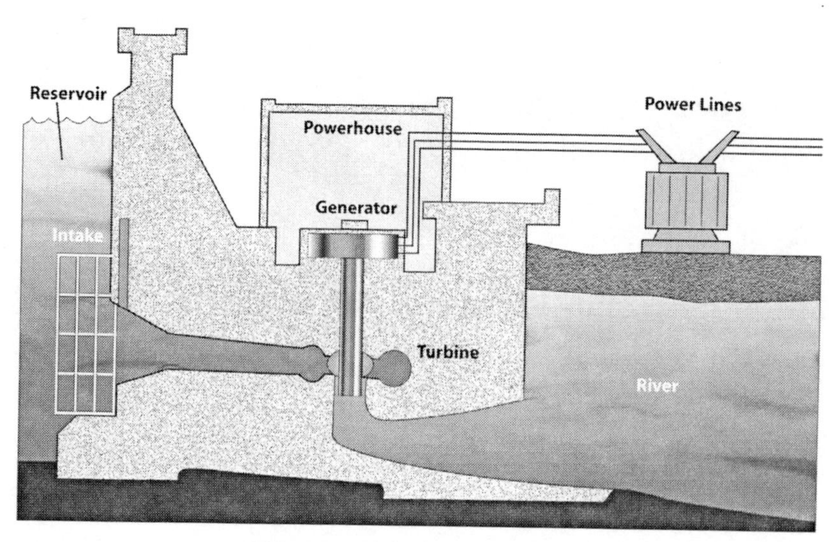

Figure 16.5 Hydroelectric Dam

Beach Reclamation

Heavy storms that cause flooding can also damage beaches and shorelines. In the same way that soil is eroded by rivers and runoff, sand can be eroded from a beach. Changing the landscape of the beach with jetties, piers or other features can also change the way currents flow. **Jetties** are solid structures made of piled stones or other material. **Piers** are open structures, supported by timber pilings, that raise a walkway above the water. Man-made structures like these sometimes have the effect of increasing the erosion of the beach and the siltation of the water. This is because water is forced to flow around the jetty or pier. Different water flow patterns change the way the waves land on the beach. Sometimes the result is unnaturally severe erosion. Other times, the reverse is true: a jetty may actually be built to change water flow patterns to *stop* erosion.

Figure 16.6 Jetty and a Pier

Beach reclamation is the process of repairing a beach area that has been eroded, damaged or destroyed. This involves dredging a large body of water, like a lake or the ocean. **Dredging** means scooping out the loose sand and earth from the bottom of the body of water. A **dredge** is a scoop attached to a crane and a barge to place the sand or mud into. The dredged material is then taken to the shoreline to restore the damaged area. Doing this allows the beach to be widened or built up in height. Another type of dredge uses pumps to actually suction sand up and shoot it out on shore, as shown in Figure 16.7.

Figure 16.7 Shooting Dredged Sand Out on the Beach

The processes that caused the damage in the first place may continue to occur. So, beach reclamation is usually an ongoing process. Dredging will have to be repeated regularly to maintain the beach.

Seismological Studies

Seismological studies are the studies of earth movement. Usually, the earth movement is caused by an earthquake. Scientists study earthquakes to learn more about the Earth. As with floods and hurricanes, the goal is to predict them. Unfortunately, this is often not possible.

The death and damage caused by seismic waves can be huge. In 1906, a powerful earthquake hit San Francisco. Figure 16.8 shows a little of the damage from that quake, and the fires that followed it.

Figure 16.8 The 1906 San Francisco Earthquake

Another example is the tsunami of December, 2004. A large earthquake in the ocean causes water to move suddenly. This creates a large wave called a **tsunami** that radiates out from the epicenter of the earthquake. This struck coastlines surrounding the Indian Ocean with 7-foot waves. It is estimated that this tsunami killed 200,000 people.

Figure 16.9 A Tsunami Wave Strikes

A seismologist uses a **seismometer** to record the size and force of a seismic wave. A simple seismometer is a pendulum with a pen attached to the end. During an earthquake, the pen stays still while the rest of the device shakes with the earthquake. The pen marks on paper that is attached a drum as in Figure 16.10. The markings indicate the direction of movement. Modern seismometers use electronic sensors hooked up to a computer to get the same effect.

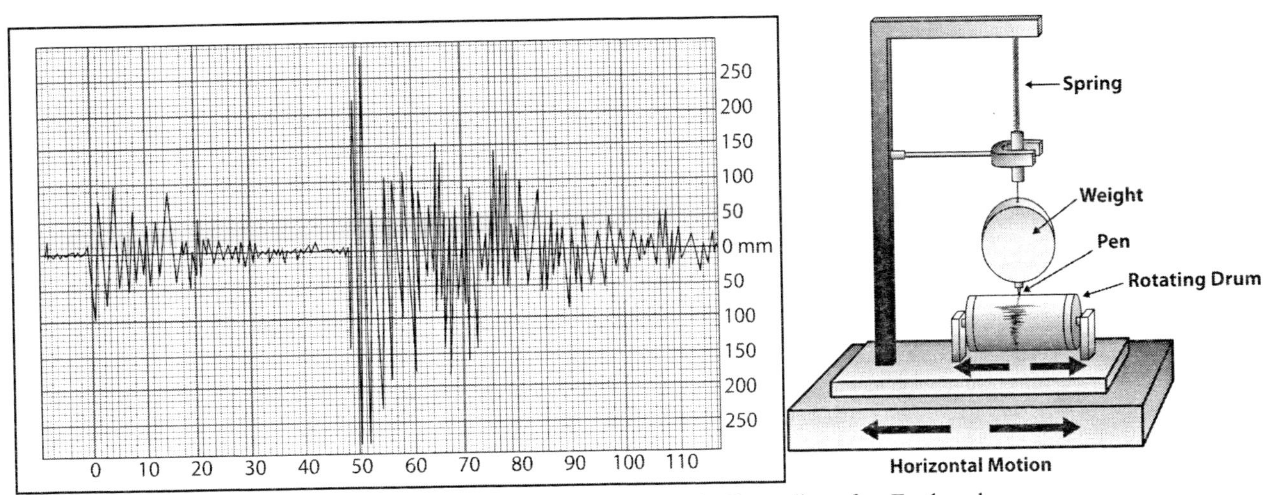

Figure 16.10 A Seismometer and a Recording of an Earthquake

Using seismometers, scientists have learned a lot about earthquakes. They have also learned about the structure of the Earth's interior. During earthquakes, seismic waves travel through the Earth's mantle and core, as well as at the crust. A network of seismometers can compare the speed of waves between two points on Earth. That way, they can figure out the speed of the waves in different places.

Why is that important? Well, seismic waves, like sound waves, are **mechanical waves**. They must travel through a **medium** (some type of matter). Rock, dirt and air are all different kinds of media. They travel faster when they are in a denser medium. **Density** describes how much mass something has relative to its size. It is related to the number of atoms packed into the space of the object. Consider a piece of popcorn and a piece of meat. Both are the same size. The meat piece is heavier than the popcorn piece, and so we say it is more dense than the popcorn.

In general, solids are denser than liquids, and liquids are denser than gases. This is not true of water. Water is more dense in the liquid state than it is in the solid state. So, if a wave travels faster in one layer of the Earth than in another, there is a difference in density between the two. In fact, this is how scientists figured out where the boundaries of the core, mantle and crust were and what they were made of.

Challenge Activity

Draw a model of the solid Earth. Label the core, mantle and crust. Label the asthe[illegible] and the lithosphere. From what we said about how seismic waves travel, label w[illegible] solid Earth seismic waves will travel slowest, and where they will travel fastes[illegible]

Chapter 16 Review

1. A reservoir is

   A. a flood.
   B. a levee.
   C. a man-made lake.
   D. a dam.

2. What do seismometers measure?

   A. seismic waves
   B. sound waves
   C. sewage drains
   D. levee strength

3. If a meteor were to land in the Pacific Ocean, what natural disasters might it cause?

   A. a hurricane and a tsunami
   B. an earthquake and a tsunami
   C. a volcano and an earthquake
   D. a volcano and a tsunami

4. A portion of Lake Lanier becomes too shallow to navigate a boat through. How could this be fixed?

   A. by building a levee
   B. by building a dam
   C. by dredging the blocked area
   D. by adding drains

5. Which of the following is the best comparison to a levee?

   A. a grate
   B. a wall
   C. a mirror
   D. a curtain

Historical Challenge Question

1. What two bodies of water overcame the levees to flood New Orleans during Hurricane Katrina?

2. Many government officials said, "The levees were not sufficient to withstand the hurricane." What did they mean? What should have been different about the levees? How would they know they were not good enough?

Building the Toomewatchee Dam

The middle Georgia town of Toomewatchee City was planning to build a new dam. The dam was to be a ____________ (adjective) thing. It would dam the ______________________________ (silliest word that you can think of) River and provide power for __________ (number) homes.

The power generated at the dam would give the Beautiful ____________ (Girl's name) Hair Salon the power to run its dryers and curling irons. It would give the _________ (Adjective) _________ (Animal) Arcade the power to run its video games. It would give the __________ (Adjective) __________ (Food) Diner the power to cook food and brew coffee.

And, while all those things were important, some people were upset about the dam. They worried about the __________ (adjective) __________ (land animals) that lived near the river. These animals were so __________________ (personality trait)! And there were only __________ (number) of them!

The worries weren't only about animals, though. The people of Toomewatchee City worried about flooding. The dam was to be close to town. What if ____________ (type of rock) rocks built up in the reservoir, and it had to be dredged? Dredging would involve the use of a ____________ (type of heavy equiptment) to ________ (verb) the _________ (adjective) rock and soil from the bottom of the reservoir. That would be expensive.

Town leaders decided to build the dam, but they were careful. First, they hired _________ (adjective) engineers to look at the plans for the dam. The engineers developed a model that said that the levee should be __________ (number) feet high and built with concrete, __________ (material) and ___________ (material). They planted ___________________ (type of plant, plural) all around the base of the levee. Clearly, this was a special levee.

Second, they formed an environmental committee. The committee decided to make an area around the reservoir as a wildlife preserve. The animals could ____________ (verb) and ____________ (verb) as much as they liked in the refuge. No one would ____________ (verb) them. The entrance to the preserve was marked with a sign. It said "Welcome to the ____________ (Adjective) ____________ (Noun) Wildlife Refuge! Your arrival makes us ____________ (adjective)!" Visitors seemed puzzled when they read the sign.

The people of Toomewatchee City finally felt that their plans were ____________ (adjective). They built the dam. And when the first rains came, no one was sitting around, waiting for the ____________ (landform) to ____________ (destructive process). No, sir! They were out playing ____________ (Verb) the ____________ (Adjective) ____________ (Food) at ____________ (Girl's Name)'s ____________ (Type of Building) of Fun!

Challenge Question

1. How did the people of Toomewatchee City prevent problems that their new dam may cause?
2. At one point in the MadLib, a ____________ (type of rock) is asked for. Ted wrote "magma". Guofeng wrote "igneous". Which answer do you think was correct?

Earth Science Domain 3 Review

1. A levee would be best used to

   A. move water off a flooded street.
   B. to dredge a river bed.
   C. to divert water from a city.
   D. to generate hydroelectric power.

2. Water damages roads in colder areas of the world. What happens to water when it freezes and thaws that causes this damage?

   A. expansion and gaining mass
   B. expansion and contraction
   C. expansion and reaction
   D. expansion and distraction

3. Which of the following is not a mode of interaction for tectonic plates?

   A. rubbing against each other
   B. rolling over each other
   C. bumping into each other
   D. pulling apart from each other

4. In which media will seismic waves go the slowest?

   A. water
   B. air
   C. igneous rock
   D. metamorphic rock

5. We work to protect humans, animals and property from the worst of Earth's natural processes. What is the first step in preventing destruction in the event of a natural disaster?
   A. predicting
   B. preparing
   C. reclaiming
   D. reconstructing

6. What was the ancient supercontinent called?
   A. Pangaea
   B. Pandora
   C. Panacea
   D. Polyanna

7. Hydroelectric power can be generated by
   A. still water.
   B. moving water.
   C. humid air.
   D. dry air.

8. Describe a constructive process.
   A. one that destroys landforms
   B. one that erodes landforms
   C. one that weathers landforms
   D. one that builds up landforms

9. In the Sahara desert, sandstorms often blow across the desert. Classify this process.
   A. wind power
   B. wind weathering
   C. wind erosion
   D. wind exposure

10. Seafloor spreading occurs in response to
   A. movement of rock in the inner core.
   B. movement of rock in the outer core.
   C. movement of rock in the mantle.
   D. movement of ocean currents.

5th Grade CRCT Science Post Test 1

1. What characteristic listed below places animals in a group separate from plants? S5L1a

   A multicellular

   B vertebrate

   C movement

   D reproduction

2. Which of the materials on the following list is the MOST electrically conducting? S5P3c

   A ink

   B shoelaces

   C salt water

   D sugar cubes

3. Why is the formation of a river delta considered a constructive process? S5E1a

   A Because it removes soil from the river bank.

   B Because the soil is moved by water.

   C Because the soil is rich in nutrients.

   D Because the soil builds up the delta.

4. The shape of your nose is S5L2a

   A a genetic trait.

   B learned behavior.

   C DNA behavior.

   D the result of protein production.

5. Why would mining be considered a destructive process? S5E1b

   A Because miners dig holes deep in the earth.

   B Because miners use tools.

   C Because miners use gasoline.

   D Because miners are looking for gold.

6. Paulie pours milk over his Sugar Crunchy Bran Flakes. Then he stirs in raisins. Then he eats the cereal. At which point is Paulie using a chemical process? S5P2b

   A pouring milk over the cereal

   B stirring in the raisins

   C digesting the cereal

   D pouring and digesting

7. If the two animals pictured here were crossed what would the babies look like? S5L2b

African Leopard African Lion

   A The babies would look like a lion.

   B The babies would look like a leopard.

   C The babies would look like a mixture of a lion and a leopard.

   D The babies would look like a German Shephard and a Chihuahua.

8. Humphrey was a Humpback Whale that twice swam into the San Francisco Bay. (Yes, it's really true!) One of those times he swam up the Sacramento River. He crossed under a bridge in a shallow stretch of the river and became stuck. Which of the following actions could have been used to help Humphrey out? S5E1c

   A reclaiming the beach along the river

   B dredging the river

   C doing seismological studies on the river bottom

   D damming the river

Go On

Use the following figure to answer questions 9 and 10.

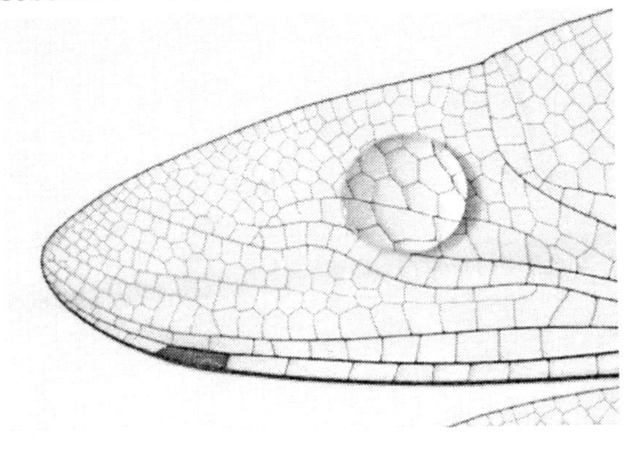

9. The picture shows how a water droplet affects the way we see a dragonfly wing. Which statement describes the effect? S5P1b

   A The water droplet has no effect on how we see the wing.

   B The water droplet magnifies about 2x.

   C The water droplet magnifies about 10x.

   D The water droplet magnifies about 40x.

10. What kind of instrument would you need to clearly see the individual cells of the wings? S5P1b, S5L3a

   A a telescope

   B a microscope

   C a sneakoscope

   D a magnifying glass

11. Which of the following is an example of chemical weathering? S5E1b

   A Acid rain slowly dissolves a limestone statue.

   B Tree roots push the concrete up in a driveway.

   C A river cuts a deep canyon in a once-shallow valley.

   D Hot water poured over ice cubes melts the ice.

12. Which characteristic below is ONLY true of amphibians? S5L1a

   A lays eggs

   B thin moist skin

   C is endothermic

   D has claws

13. Annie rubs a balloon against her wool sweater. Mary rubs a balloon against her gold bangle bracelet. Teresa rubs a balloon against her can of soda. Each of the girls then holds their balloon against the wall. Which one sticks the longest? S5P3a, c

   A Annie's

   B Mary's

   C Teresa's

   D None of the girls' balloons

14. Identify the type of cell shown below. S5L3a

A plant cell

B animal cell

C tree cell

D flower cell

15. Identify the true statement below about microorganisms. S5L4a, b

A The size of an organism determines its importance in an environment.

B Tiny microorganisms cannot harm or kill larger organisms.

C Humans are unaffected by microorganisms.

D Microorganisms affect every living thing on Earth.

16. A handheld magnifying glass is BEST used to view which of the following items? S5P1b

A a wart on your heel

B the wings of a gnat

C a grain of table salt

D an eyelash

17. Which pair of lists below groups animals into birds and mammals? S5L1a

A

salamander	iguana
frog	turtle
newt	snake

B

iguana	blue whale
dove	shark
parrot	squid

C

blue whale	finch
mouse	ostrich
kangaroo	hawk

D

humans	lizard
butterflies	turtle
frogs	snake

Go On

18. Sean has a purple piece of construction paper on his desk. His teacher asks him to name an activity that would chemically change the paper. Which of his answers is correct? S5P2c, a

A Pour water on the paper.

B Squish the paper into a ball.

C Eat the paper.

D Put magnets on the paper.

19. Mid-ocean ridges are formed under oceanic crust. Material flows up through cracks in the crust and cools to form new crust. What material flows up through the cracks? S5E1a

A marmalade

B metamorphic rock

C magma

D magnesium

20. A wastewater treatment plant removes several types of contamination from a particular stream. Which of the following is NOT removed by filtration? S5P2c, a

A plastic bags

B sticks

C sand

D dissolved lead

21. Which of the following items could you pick up with a magnet? S5P3d

A a steel can

B a roll of toilet paper

C a water bottle

D a diamond

22. Which organism listed below is an invertebrate found in your yard? S5L1a

A squid

B clam

C dog

D earthworm

Use the following figures to answer questions 23 – 24.

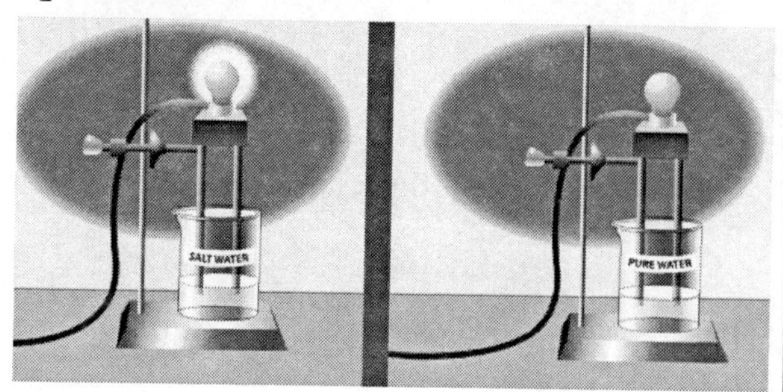

23. What conclusion can you reach, given the figure above? S5P3c

    A Salt water is a better electrical conductor than pure water.

    B Salt water is a better electrical insulator than pure water.

    C Light bulbs are powered by water.

    D Light bulbs need gasoline to turn on.

24. There is a black cord connected to the light bulb in each figure. What must the other end of the black cord connect to? S5P3b

    A another light bulb

    B a resistor

    C a power source

    D an open switch

25. Mountain ranges tend to form in long lines or arcs. These linear ranges mark the location of S5E1a

    A a string of islands.

    B a volcano.

    C a tectonic plate boundary.

    D oceanic crust.

26. How do bacteria and fungi help the environment? S5L4a

    A They help us produce proteins.

    B They cause "red tides."

    C They make bread and cheese.

    D They break down dead plants and animals.

27. There is a path between Dean's house and Rick's house. The path was formed over many years, as Dean and Rick walked along the path to visit each other. Classify the process that formed the path. S5E1b

    A constructive

    B destructive

    C divergent

    D transform

28. Why do scientists use taxonomy? S5L1b
    A to confuse old people
    B to confuse students
    C to clearly identify things
    D to correct mistakes of other scientists

29. When people say "please" and "thank-you" at the dinner table, what are they showing? S5L2a
    A a genetic trait
    B learned behavior
    C DNA behavior
    D the result of protein production

30. Tip's Garage uses an electromagnet to move large items. Which of the following items is the electromagnet able to pick up? S5P3c, d
    A car engines
    B glass windshields
    C rubber tires
    D plastic bumpers

31. How are genes passed from parents to children? S5L2b
    A through traits
    B through heredity
    C through learned behaviors
    D through viruses

32. Ike uses a razor blade to slice through a piece of Styrofoam™, a piece of wood and a piece of plastic. Which of these items chemically changed as a result of his actions? S5P2a, c
    A the Styrofoam™
    B the wood
    C the plastic
    D None of the items changed chemically.

33. Which example below is a learned behavior? S5L2a
    A the length of your fingers
    B getting the flu
    C running a race
    D blinking

34. Which rock group forms from the hardening of magma? S5E1a
    A sedimentary
    B metamorphic
    C obsidian
    D igneous

35. Which item below is NOT part of an animal cell? S5L2b
    A cell membrane
    B cytoplasm
    C chloroplast
    D mitochondria

36. Which process represents a chemical change? S5P2c

A melting of ice

B corrosion of iron

C evaporation of alcohol

D crystallization of sugar

37. A volcano spews hot ash and lava that burns down houses, trees and anything else in its path. Why is the eruption of a volcano viewed as a constructive process? S5E1a

A It adds minerals to the soil.

B It adds minerals to the atmosphere.

C It takes water from the soil.

D It builds up rock as lava cools.

38. How does a cell wall help a plant cell? S5L3b

A It provides shape and support.

B It stores DNA.

C It regulates temperature.

D It makes proteins.

39. Tony boils water. Timo boils orange juice. Tamika boils milk. A vapor rises from each pan. What is it? S5P2a

A orange juice

B air

C steam

D plastic

40. How is a blood cell physically different from a nerve cell? S5L3c

A A blood cell is long and branching and has no nucleus.

B A blood cell has no nucleus and is round.

C A blood cell is the exact same as a nerve cell.

D A blood cell carries oxygen and a nerve cell carries messages.

Go On

41. Santos has a nail, electrical tape, a wire, a battery, paperclips and a bar magnet. His goal is to make an electromagnet. Which of the items mentioned will he have left over after the electromagnet is complete? S5P3d

A the nail, the wire and the battery

B the nail and the paperclips

C the battery and the paper clips

D the paperclips and the bar magnet

42. Geologists measure earthquakes using a seismometer. What does a seismometer measure? S5E1c

A mass of the soil moved during the earthquake

B distance that the seismometer moved during the earthquake

C changing weather conditions during the earthquake

D heat of the earth at the epicenter of an earthquake

Use the following figure to answer questions 43 and 44.

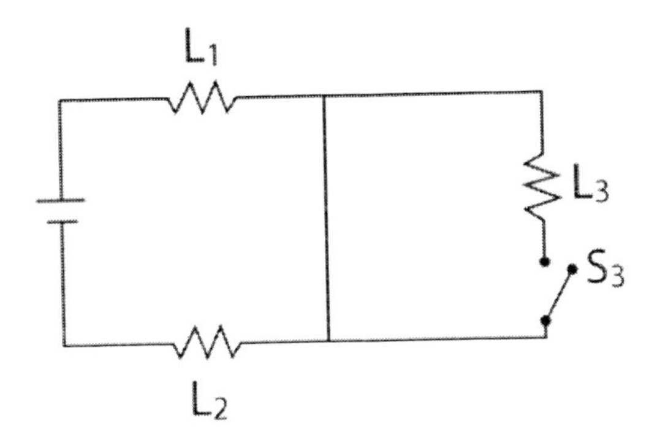

43. How many open switches are shown in the electrical circuit? S5P3b

A 0

B 1

C 3

D 4

44. What material should the circuit elements be connected with? S5P3c

A shoelaces

B a chain of paper clips

C copper wire

D ice

45. What item listed below do plant cells use to make food? S5L3b

A protein

B cotton

C sunlight

D green coloration

46. Construction adds to erosion by S5E1b

A creating areas of drought.

B adding soil to the area.

C removing plant cover from the soil.

D adding plant cover to the soil.

47. Identify a way that bacteria do NOT enter the human body. S5L4a, b

A under the finger nails

B through the stomach

C through the lungs

D cut in the skin

48. Bob, the class hamster, weighed 155 grams. Over the holidays, Bob did not get brushed every day, and his long hair got badly tangled. When the class returned, they helped their teacher shave Bob. They also clipped his long nails. How much might Bob have weighed after this grooming? S5P1a

A 50 grams

B 149 grams

C 155 grams

D 185 grams

49. Vegetable food scraps can be placed in an outdoor compost pile to decompose. Which of the following is reasonable evidence that a chemical reaction is happening in the compost pile? S5P2c

A The pile gets warmer.

B The pile disappears.

C The pile gets wet when it rains.

D Raccoons eat the food scraps.

50. A can of cream of celery soup has a mass of 305 grams. If the 5 gram lid is removed and 250 grams of soup is poured into a pan, what is the mass of the now empty can? S5P1a

A 200 grams

B 50 grams

C 55 grams

D 65 grams

51. Which of the following landforms is NOT located in the Pacific Ring of Fire? S5E1a, b

A the Mayon volcano, Philippines

B Mt. St. Helens, USA

C Stone Mountain, USA

D the Mount Merapi volcano, Indonesia

Go On

52. Copper is a light, shiny metal that conducts heat and electricity well. Which of the following activities will chemically change a 1 kilogram block of copper into something else? S5P2a, c

A melting it

B hammering it

C pouring acid on it

D painting it

53. Some unicellular protists can be seen by the naked human eye. Some multicellular organisms can only be seen with a microscope. What does this tell you? S5L3c

A Cells come in only one size.

B Unicellular organisms fill themselves with air.

C Cells come many sizes and can be quite big.

D Unicellular protists are not alive.

54. Sugar is placed in a glass jar. Sulfuric acid is poured on top. The sugar turns black and expands, rising into a tall column above the jar. What has happened? S5P2c

A The sugar has chemically changed into something else.

B The sugar is still sugar and has only changed physically.

C The sulfuric acid changed air into sugar.

D The sugar was eaten by the sulfuric acid.

55. Cell organelles are suspended in a material inside the cell. What material is that? S5L3b

A water

B acid

C cytoplasm

D chloroplasm

56. What can a bar magnet do that an electromagnet cannot? S5P2c

A Pick up metal objects.

B Pick up all objects.

C Work without being powered by a battery.

D Make chemical reactions occur.

57. Mr. Biolo's class wanted to go to the new aquarium exhibit, Invertebrate World. He asked four class members what they wanted to see at the exhibit. Which student would be unable to see their chosen animal at the Invertebrate World exhibit? S5L1a

A Samar, who asked to see the squid.

B Dean, who wanted to see the sharks.

C Marlo, who asked to see the jellyfish.

D D'Angelo, who wanted to see the lobsters.

58. How do viruses cause disease? S5L4b

A by putting their own DNA into human cells

B by producing toxins that damage tissues

C by causing red tides that contaminate food and water

D with vaccines

59. The flowering plants are a group of plants that have over 250,000 different species, more than all the other plant groups combined. What characteristic MOST likely made this group of plants so successful? S5L1b

A using non-vascular tissue to transport water

B using flowers to make seeds

C having hearty roots to reach deep into soil

D having green color to make glucose

60. Black mold is an organism that can be very dangerous to humans. Spores inhaled from this mold can kill infants, children and the elderly. This type of organism likes to live in dark, damp places like basements. What type of microbe is this organism? S5L4b

A bacteria

B fungi

C protists

D virus

61. What is a bunch of muscle cells grouped together called? S5L3c

A stomach

B lung

C muscle tissue

D muscular system

Go On

62. Why is heredity important? S5L2b

A It passes traits from parents to offspring.

B It makes the DNA inside cells similar.

C It turns DNA into proteins the cells can use.

D It creates cell walls.

63. Common marram grass is often planted by humans on beach dunes. Which of the following would be the BEST reason for this intentional planting? S5E1b

A to increase dune erosion

B to decrease dune erosion

C to increase sand weathering

D to decrease sand weathering

64. What type of disease can you get from a virus? S5L4b

A cavities

B chicken pox

C food poisoning

D athlete's foot

65. Why does static electricity MOST often occur when the weather is dry? S5P3a

A Because electricity needs water to travel.

B Because electrons are destroyed by water.

C Because water in the air carries off the charge before it builds up.

D Because water is an insulator and wraps up the charge.

66. When an egg is cooked, it goes from runny liquid to a firm solid. When water is frozen, it goes from a flowing liquid to a hard solid. Which of these is a chemical process? S5P2a, b, c

A Cooking an egg is chemical process.

B Freezing water is a chemical process.

C Neither action is a chemical process.

D Both are chemical processes.

67. A lion and a tiger were crossed. Select the picture that MOST resembles what the baby would look like when it grew up. S5L2b

A 

B 

C 

D 

68. What cell part below is found on the outside of plant cells? S5L3b

A cell membrane

B nucleus

C chloroplasts

D cell wall

69. To follow are four Georgia landforms. Which has the least need for structures to control flooding? S5E1c

A Okeefenokee Swamp

B Brasstown Bald

C The Chattahoochee River

D The Georgia Barrier Islands

70. Francois finds a lustrous, shiny rock at the edge of the park. He places it in his pocket. When he takes the rock out later, he finds that the change in his pocket is stuck to the rock. What is Francois' rock? S5P3d

A marble, a metamorphic rock

B magnetite, a mineral of iron

C silver metal

D steel, an alloy of iron and carbon

STOP

5th Grade CRCT Science Post Test 2

1. Maya has a box of chicken nuggets. The package label says that each nugget has a mass of between 6 and 10 grams. What is the maximum mass of chicken nuggets that Maya could eat, if she ate 4 nuggets? S5P1a

   A 16 **B** 24 **C** 40 **D** 60

2. What happens when a substance melts? S5P2b

   A It disappears.

   B It changes into a solid.

   C It changes into a liquid.

   D It changes into another kind of substance.

3. What characteristic below separates plants into different groups? S5L1b

   A having a backbone

   B endothermic body type

   C having vascular tissue

   D laying eggs

4. A jagged chunk of granite stands at the edge of an old granite quarry. It is 3 meters high today. How tall could it have been when the quarry closed 30 years ago? S5E1b

   A 1 m **B** 2 m **C** 3 m **D** 4 m

5. A microorganism gets inhaled through the nose. This microbe travels down the trachea and into the lungs. Once inside the lungs, this microbe inserts its DNA into the lung cells and begins creating more microbes. What type of microbe is this? S5L4b

   A bacteria **B** fungi **C** protist **D** virus

6. A levee is used to S5E1c
   - **A** absorb water.
   - **B** re-direct water.
   - **C** chemically react with water.
   - **D** change the state of water.

Use the table and information below to answer questions 7 and 8.

Four houses in the same neighborhood have ivy growing on them. Each house is made of a different material. A building inspector carefully removes a segment of ivy from each house and measures how deep the root has grown into the material.

Material	Root Depth (centimeters)
Brick	2 cm
Wood	4 cm
Stone	1 cm
Concrete	3 cm

7. Which type of building material is the MOST weathered by the ivy? S5E1b
   - **A** brick
   - **B** wood
   - **C** stone
   - **D** cement

8. Which of the following questions would help the building inspector compare the building materials with one another? S5E1b
   - **A** Are these houses all the same size?
   - **B** Are these houses all the same age?
   - **C** Are these houses all the same color?
   - **D** Are these houses all the same price?

9. Which organism listed below is a vertebrate found at an aquarium? S5L1a
   - **A** starfish
   - **B** octopus
   - **C** penguin
   - **D** shrimp

10. Which of the following could NOT be picked up by an electromagnet? S5P3d
    - **A** iron filings
    - **B** dirt
    - **C** paper clips
    - **D** Hot Wheels™ cars

11. Which microorganisms listed below can be helpful? S5L4a
    - **A** bacteria
    - **B** fungi
    - **C** protists
    - **D** all of the above

12. A light bulb is connected to a battery by a wire. The wire is then connected back to the battery to complete the circuit. Describe how electricity flows in this circuit. S5P3b

    A from the battery to the bulb and then back to the battery

    B from the bulb to the battery and then back to the bulb

    C only inside the battery

    D from the battery to the bulb, where it stays

13. Which of the following layers of the Earth is NOT solid? S5E1c

    A oceanic crust

    B continental crust

    C outer core

    D inner core

14. What is water vapor? S5P2b

    A a solid

    B a liquid

    C a gas

    D an atom

15. Liverworts and hornworts remain small because they do NOT have what? S5L1b

    A vertebrate

    B endothermic body type

    C vascular tissue

    D non-vascular tissue

16. Invertebrates do NOT have what physical structure listed below? S5L1a

    A eyes

    B brain

    C spine

    D hair

17. Which of the following observations MUST be described with a number? S5P2c

    A color of the sky

    B smell of a vapor

    C melting point of salt

    D taste of a pretzel

18. Which of the following instruments is necessary to examine vibrations in the Earth? S5E1c

    A a seismometer

    B a tsunami

    C a pen

    D a dredge

19. How is the length of your finger determined? S5L2b

A from your DNA

B from your education level

C from your reading skill

D from your ability to tie your shoes

20. A handheld magnifying glass would be useful for observing S5P1b

A the skin of a snake.

B the motion of atoms.

C the legs of a flea.

D blood cells.

21. What type of food listed below do humans NOT use bacteria to make? S5L4a

A bread

B cheese

C yogurt

D chicken soup

22. Increasing the current applied to an electromagnet will affect which of the following? S5P3d

A the size of the electromagnet

B the magnetic field of the electromagnet

C the mass of the electromagnet

D the magnification of the electromagnet

23. Why is heredity important to living things? S5L2b

A it passes along traits to future generations

B it keeps everything looking similar

C it allows organisms to have babies

D it allows man to domesticate animals

24. Andres wants to connect the pieces of his electrical circuit. What material will make the BEST wire? S5P3c

A copper

B plastic

C glass

D dental floss

Go On

25. Which example below is a learned behavior? S5L2a

A breathing

B the shape of your eyes

C dialing the phone

D length of your legs

26. If you were to discover a rock that appeared to have layers, what kind of rock would you suspect it to be? S5E1a,b,c

A metamorphic

B igneous

C sedimentary

D lava

27. A solution is made by dissolving 10 g of salt in 500 g of water. Identify the mass of the resulting solution. S5P1a

A 500 g

B more than 500 g but less than 510 g

C 510 g

D more than 510 g

28. To demagnetize a bar magnet, you MUST S5P3d

A stop the flow of current.

B turn the battery around.

C heat or drop the magnet.

D break it in half.

29. How does a cell membrane help an animal cell? S5L3b

A it keeps cell parts inside

B it stores heredity information

C it dissolves nucleic acids

D it makes proteins

30. Which of the following BEST describes the movement of a tectonic plate? S5E1a,b

A faster than a race car

B faster than a horse, but slower than a race car

C slower than a horse, but faster than a turtle

D slower than a turtle

31. The mayor orders that a river be dredged. What will the result be? S5E1c

A The river will be deeper.

B The river will be shallower.

C The river will be cleaner.

D The river will have no more fish.

32. Tyrell mixes 75 grams of sand with 10 grams of salt. What has he done? S5P2c

A He started a chemical reaction between sand and salt.

B He made a mixture of sand and salt.

C He destroyed 10 grams of sand, and now has only 65 grams left.

D He created 85 grams of sodium gas.

33. The number of years it takes for running water to erode a canyon the size of the Grand Canyon should be measured in S5E1b

A hundreds of years.

B thousands of years.

C millions of years.

D billions of years.

34. You are an electrical worker, assigned to rewire a household circuit. Which of the following is the BEST material to cover your hands with? S5P3c

A rubber gloves

B cotton gloves

C aluminum foil

D wool gloves

35. Why do you look similar to other people in your family? S5L2a, b

A because you share similar DNA

B because you were raised in the same house

C because you learned how to look from your parents

D because you all wear similar jeans

STOP

36. After a chemical change takes place, how can the products be changed back to the way they were before the reaction? S5P2c

A They cannot be changed back.

B Another chemical reaction could change them back.

C They could be changed back by filtering.

D They could be changed back by stirring.

37. Which item below do plant and animal cells have in common? S5L3a

A cell wall

B nucleus

C flagella

D chloroplasts

38. How are protists harmful to the environment? S5L4b

A They create red tides.

B They decompose dead plants and animals.

C They cause athlete's foot.

D The give humans chicken pox.

39. If the two animals pictured here were crossed, what would the babies look like? S5L2a, b

A The babies would look like a horse.

B The babies would look like a donkey.

C The babies would look like a mixture of a horse and a donkey.

D The babies would look like a German Shepherd and a Chihuahua.

40. Tamara is asked to design a circuit to light up her classroom. Which of the following is NOT necessary to complete this electrical circuit? S5P3b

A conducting wire

B insulating wire

C power source

D light bulbs

41. Which cell part makes glucose? S5L3b

A mitochondria

B nucleus

C cytoplasm

D chloroplasts

42. A delta is formed by which of the following materials? S5E1a

A sediment

B lava

C a glacier

D magma

43. What is the function of a resistor in an electrical circuit? S5P3b

A to carry electrical current

B to create electrical current

C to resist the flow of electrical current

D to stop electrical current

44. A seismologist studies S5E1c

A dams and levees.

B earthquakes.

C coastlines.

D fossils.

45. Timothy freezes his can of soda. Then he removes it from the freezer and places it on the counter. He lets the can sit for a day, and then opens it. Which of the following is a likely entry in his lab notebook? S5P2b

A The can contains only gas.

B The can contains solid soda.

C The can contains liquid soda.

D The can contains a new kind of soda.

46. What organism listed below is a vertebrate found in a freshwater lake? S5L1a

A seal

B blue whale

C bass

D jellyfish

47. What do you call a bunch of bone cells grouped together? S5L3c

A cartilage

B bone tissue

C skeleton

D tendon

Go On

48. An apple left at the bottom of a bookbag for several days undergoes a chemical change. Which is it? S5P2c

   A change of state

   B bubbling

   C rotting

   D burning

49. Where might you find a group of people planning a series of beach reclamation activities? S5E1c

   A Tallulah Falls

   B Jekyll Island

   C Brasstown Bald

   D Stone Mountain

50. Identify a statement that shows how bacteria are NOT helpful to humans. S5L4b

   A Bacteria break down dead plants and animals.

   B Bacteria break down water, soil and air pollution.

   C Bacteria cause inflammation of tissues and often fever.

   D Bacteria break down the milk protein to create cheese.

51. What would be the BEST magnification to use to observe a 1 millimeter long insect under the microscope? S5P1b

   A 2x

   B 5x

   C 40x

   D 1000x

52. The ability to read is a/an S5L2a

   A genetic trait.

   B learned behavior.

   C DNA behavior.

   D result of protein production.

53. What do chloroplasts do? S5L3b

   A break down food

   B capture energy and make food

   C provide support for the cell

   D make protein

54. Which of the following materials is LEAST likely to build up static electricity? S5P3a,c

   A rubber

   B wool

   C aluminum

   D plastic

Go On

55. Which of the following is a characteristic of a fish but NOT of a reptile? S5L1a

A They lay eggs.

B They have scales.

C They have gills.

D They have hair.

56. The poles of a bar magnet are found S5P3d

A at either end of the magnet.

B in the center of the magnet.

C only when the magnet touches another magnet.

D when you allow current to flow.

57. Which of the following is NOT a vertebrate? S5L1a

A a worm

B a bird

C a salamander

D a snake

58. A sculptor makes a statue from a block of clay. Under which condition would the mass of the sculpture be equal to the mass of the block of clay? S5P1a

A The sculptor worked slowly.

B The day was very hot.

C The final sculpture contains all the original clay.

D A chemical reaction took place.

59. A nerve cell looks different from a blood cell because S5L3c

A it has a different function.

B it is made from a different organism.

C it transports proteins.

D it has different DNA.

60. An earthquake is the result of S5E1b

A erosion.

B tectonic plate movement.

C human intervention.

D seismological studies.

61. Which microorganisms cause malaria? S5L4b

A bacteria

B fungi

C protists

D viruses

Go On

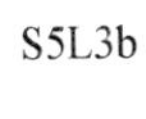

62. Miguel adds 100 grams of sugar to 100 grams of wood shavings. What could he do next, if he wants to cause a chemical reaction? S5P2c

    A ignite the mixture

    B wait for the mixture

    C stir the mixture

    D add more sugar to the mixture

63. Moss growing on the wall of a barn is an example of S5E1b

    A erosion.

    B weathering.

    C tectonic movement.

    D technology.

64. Which example below is NOT an inherited trait? S5L3a

    A thick, curly hair

    B long, pointed nose

    C riding a bike

    D short arms

65. Which of the following organelles is the cell membrane? S5L3b

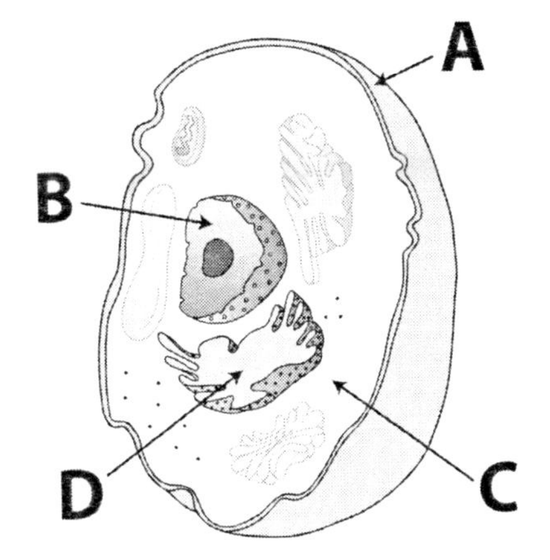

    A organelle A

    B organelle B

    C organelle C

    D organelle D

66. A constructive process results in S5E1a

    A the build-up of landforms.

    B the destruction of landforms.

    C the build-up of tectonic plates.

    D the destruction of tectonic plates.

67. Which grouping separates plants into vascular and non-vascular? S5L1a

A

snake	carrot
goldfish	orchid
finch	pine

B

maple tree	celery
sunflower	bacteria
tomato	alga

C

carrot	moss
pine tree	liverwort
orchid	hornwort

D

tortoise	squid
redwood	potato
plankton	fern

68. Which of the following actions produces friction? S5P3a

A sliding on ice

B reading a book

C floating in space

D adding two numbers

69. Mr. Tanaka's backyard gets waterlogged when it rains. It stays soggy for days after the rain has stopped. What would be the BEST way to stop this from happening? S5E1c

A Put a pool in the yard.

B Dredge the yard.

C Put drains in the yard.

D Perform seismological studies on the yard.

70. Which component of an electrical circuit would you use to turn off an electromagnet? S5P3b, d

A a power source

B wire

C a switch

D a resistor

STOP

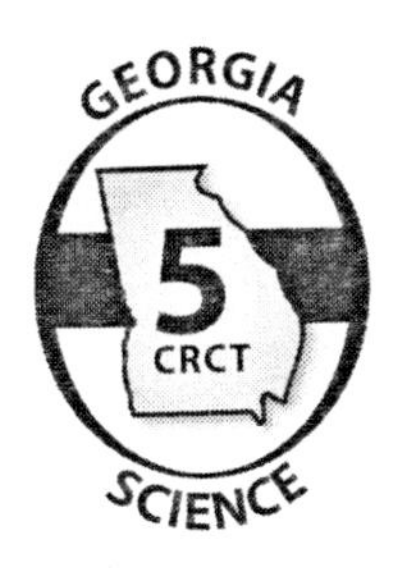

Appendix A

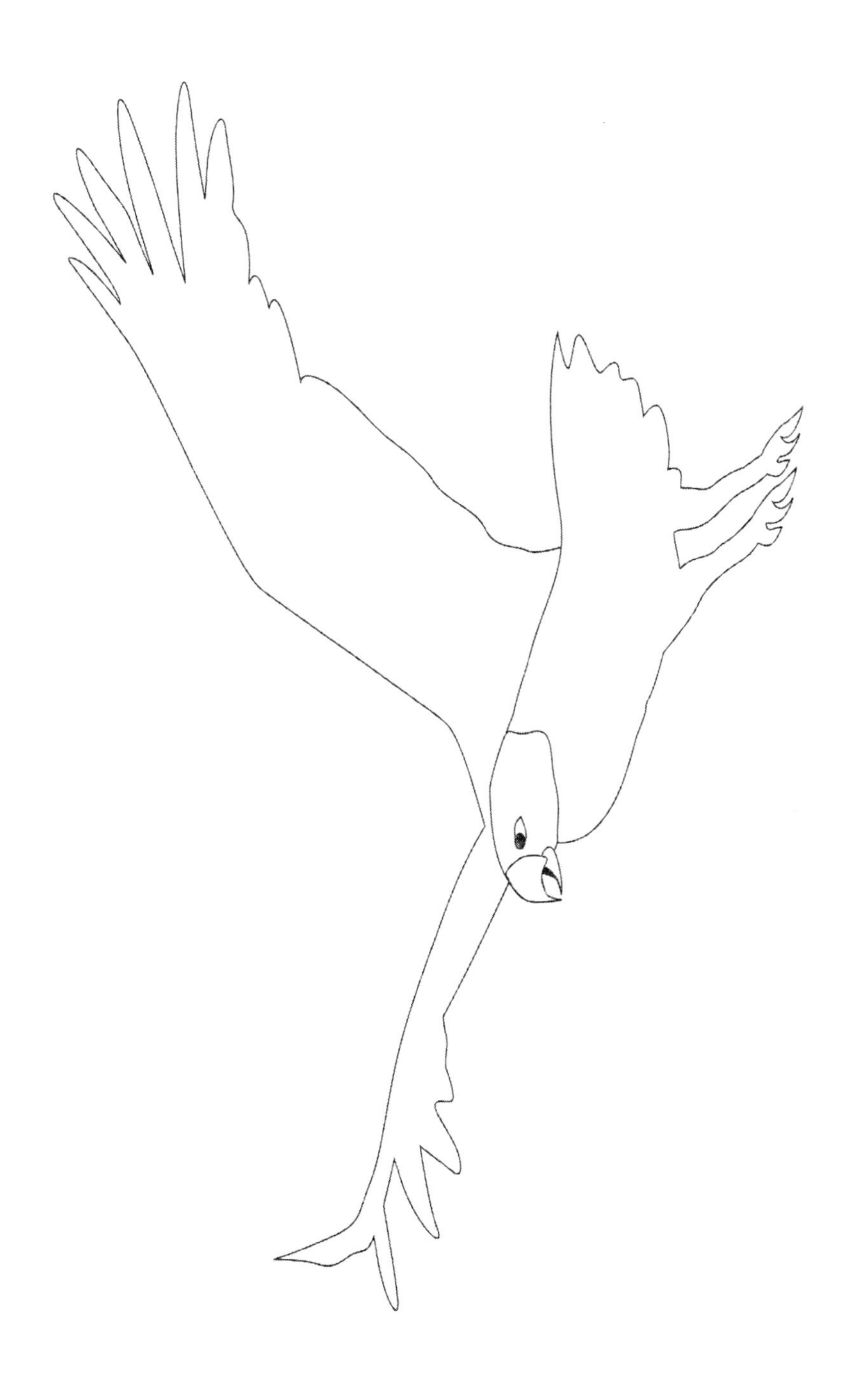

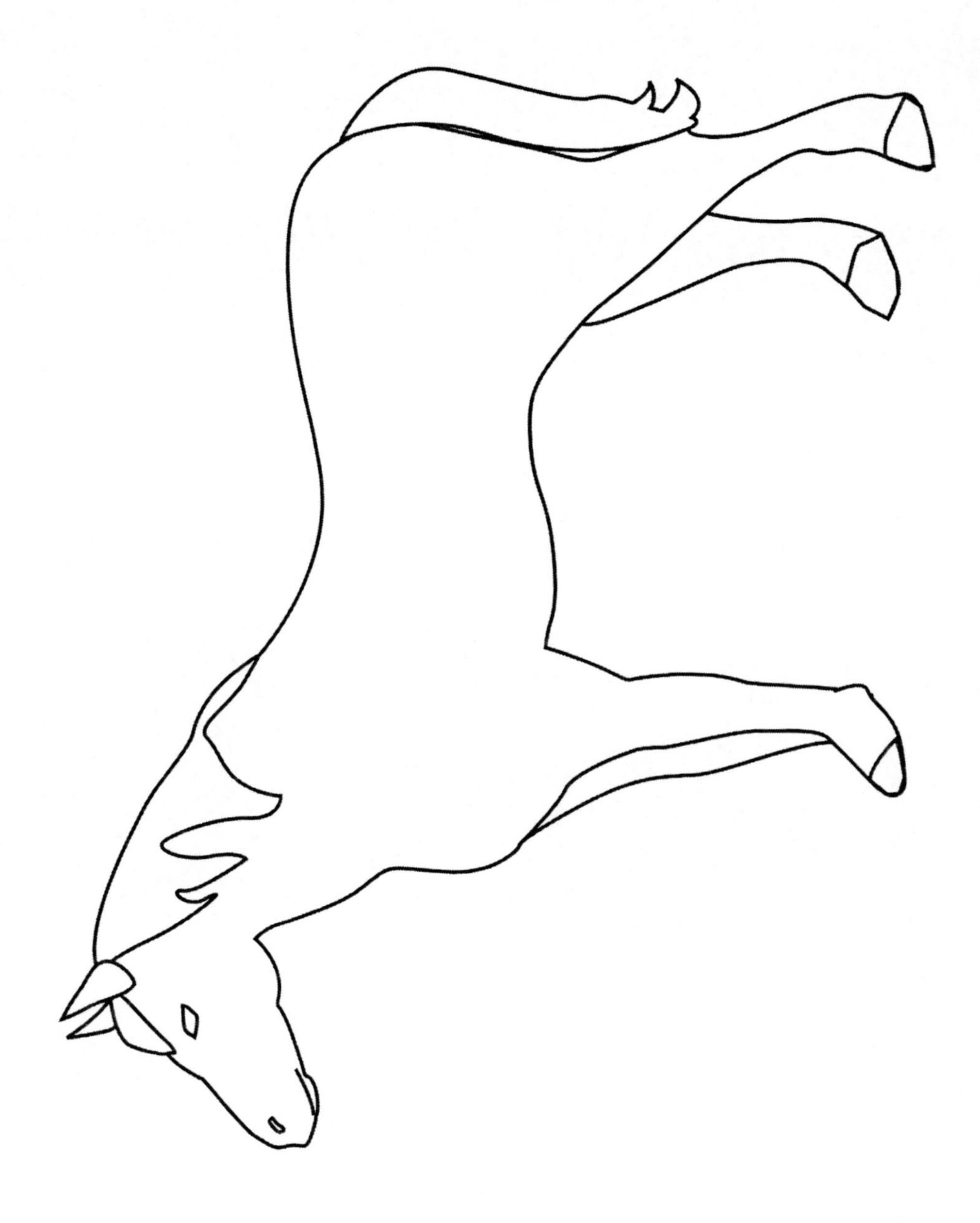

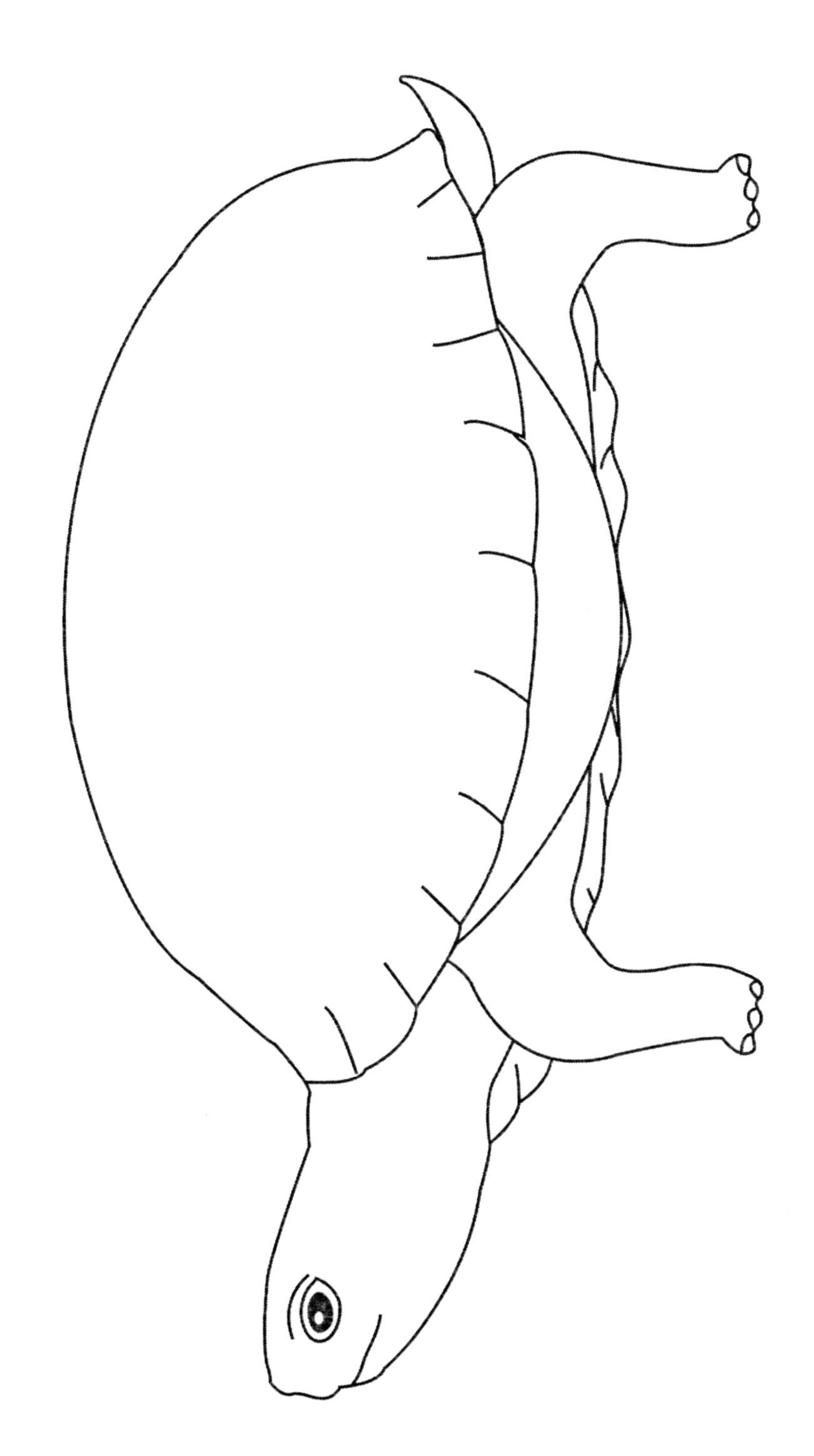

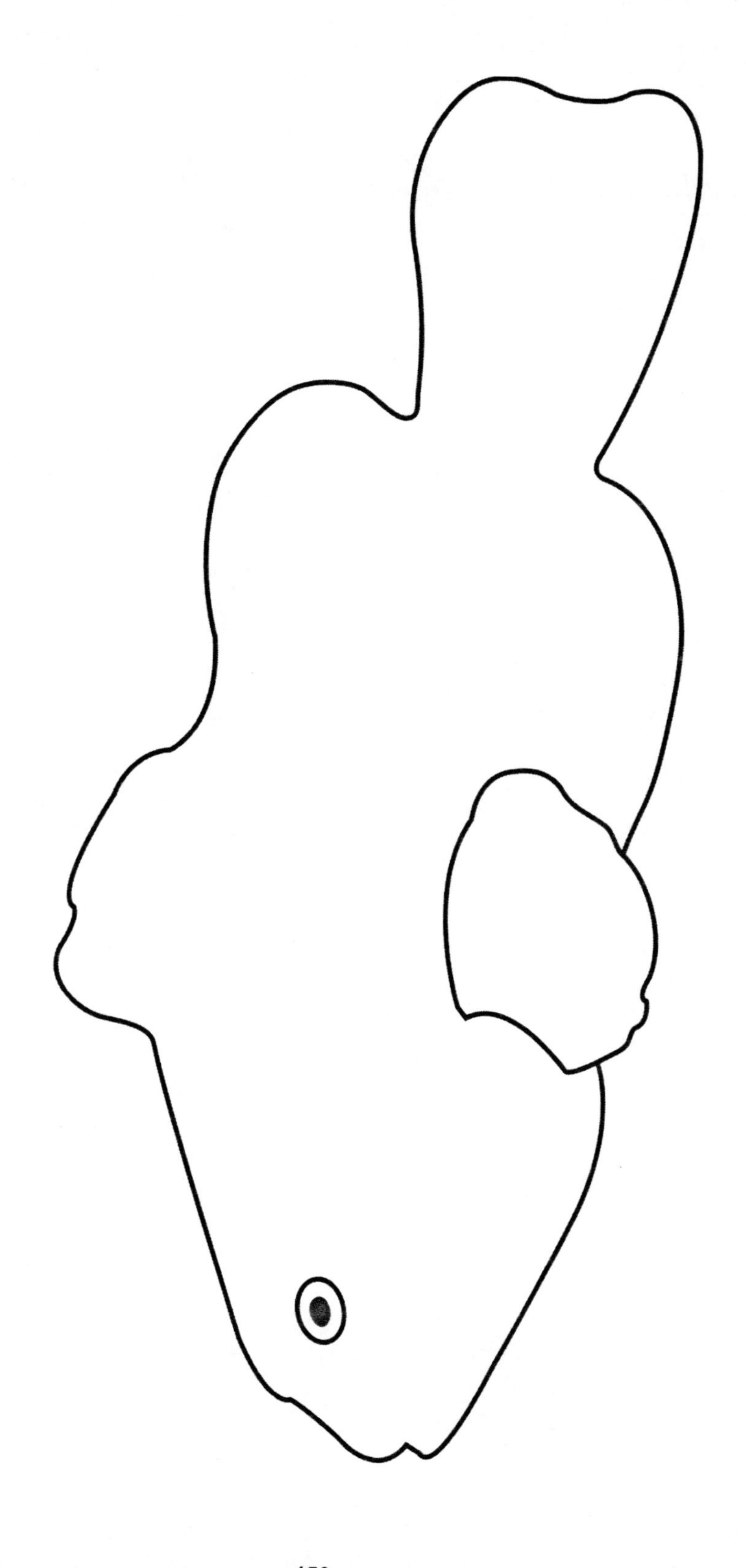

A

amphibian 71
asthenosphere 100
atoms 28
attract 53

B

bacteria 90
batteries 50
beach reclamation 117
beak 72
biological classification 65
biological weathering 105, 106
bird 72
Brown, Robert 28
Brownian Motion 28

C

cartilage 71
chemical reactions 39
chemical weathering 105, 106
classification 65
claw 72
compound 38, 39
computer model 114
constructive process 102, 107
continental crust 99
continental drift 101
core 100
crater 108
crust 99

D

dam 115
deep sea trench 102
density 119
deposition 107
destructive process 102, 107
dissolved 33
dredge 117
dry skin 72
dust bowl 107

E

Einstein 28
electrical circuit 50
electrical conductors 45
electrical insulators 45
electricity 45
electromagnet 55
electrons 28, 45
endothermic 72
erosion 106
  types of 106
exothermic 71

F

feather 72
filter paper 34
fin 71
fish 71
floodplains 114
floods 113
four legs 72
fungi 91

G

gene 85
gill 71
Guyot 103

H

harmful microorganisms 89
helpful microorganisms 89
heredity 85
heterogeneous 34
homogeneous mixture 33
hydroelectric 115

I

igneous 105
igneous rock 109
inherited traits 85
internal skeleton 70
invertebrate 70
island 103

J

jetties 116

L

landform 102
lava 108
Law of Conservation of Matter 22, 39
lay eggs 72
learned behavior 86
levee 114
light microscope 26
lithosphere 99

M

magma 101, 108, 109
magnet, de-magnetize 55
magnetic field 53
magnification 25
magnifying glass 25
mammal 73
mantle 99
mass 21
matter 21
mechanical waves 119
mechanical weathering 105
medium 119
metamorphic 105
metamorphic rock 109
metric unit 21
microorganisms 89
microscope 25
mid-ocean ridge 101
mixture 33, 38
molecule 39
molten 99

multicellular 70

N

neutrons 28
non-vascular 75

O

objective 25
oceanic crust 99
optical microscope 26

P

Pangaea 101
perpendicular 56
phloem 75
physical changes 32
piers 116
plate boundary 108
plate tectonic 100
property 21
protists 90
protons 28

Q

qualitative description 21
quantitative description 21

R

red algae 90
repel 53
reptile 72
reservoir 114
resistor 50
Rift Valley 103
rock cycle 109

S

scale 71
scientific name 67
seafloor spreading 101
Seamount 103
sediment 105, 109
sedimentary 105
sedimentary rock 109
seed 76
seeded vascular plant 76
seedless vascular plant 76
seismological studies 117
seismometer 118
sensitive skin 72
sewage system 114
solution 33
species 66
storm drainage system 113

T

taxonomy 65
tissue 75
trait 85
tsunami 106, 118

U

unit 21

V

vaccine 90
vascular 75
vent 108
vertebrate 70
viruses 90

W

water erosion 106
weather 105
  types of 105
weathering 105

X

xylem 75